MATHEMATICS
for
BIOLOGISTS

MATHEMATICS

for

BIOLOGISTS

Alan Crowe

Department of Zoology,
University of Durham, Durham,
England

and

Angela Crowe

1969

ACADEMIC PRESS

LONDON · NEW YORK

ACADEMIC PRESS INC. (LONDON) LTD
24-28 Oval Road
London NW1

U. S. Edition published by
ACADEMIC PRESS INC.
111 Fifth Avenue
New York, New York 10003

Library of Congress Catalog Card Number: 73-92401
Standard Book Number: 12-198250-5

Printed in Great Britain by offset lithography by
Billing & Sons Limited, Guildford and London

PREFACE

A glance through any modern textbook or journal in the fields of bio-chemistry, ecology, genetics or physiology reveals an ever increasing use of mathematics which might range from the solution of complicated differential equations in population studies on the one hand to the use of transfer functions in the analysis of eye-tracking mechanisms on the other. Most undergraduates reading biology or medicine have at school concentrated on a combination of subjects for which the minimum of mathematics is required; they therefore find themselves at a disadvantage at University. Several mathematics textbooks have been written for biologists but these usually deter the student: their scope is so great that the student can neither select nor anticipate his needs, or the standard is too advanced for him to comprehend, or it is difficult for him to see the relevance of the material because it is not backed up by biologically orientated examples.

In this book we have attempted to provide a comprehensive course for those biologists who have had insufficient formal training in mathematics. Most of the material does not extend beyond that which will be found in the advanced-level school mathematics syllabus, and some of this syllabus, such as trigonometry or solid geometry is not included. Very little statistics is covered since there are so many excellent texts at this level available. The pattern of the book is such that, as far as possible, each topic is backed up by exercises for the student to work out for himself. It cannot be emphasised too strongly that as much *regular* time as possible be given to working out examples, not only in this book but also those contained in the relevant sections of advanced level mathematics text-books.

No claim is made that this book goes to the highest standards of mathematics that a particular biologist will require, but it is hoped that it will be adequate to enable him to progress to more advanced texts that cater for his own speciality.

June 1969

ANGELA CROWE

ALAN CROWE

CONTENTS

4. Non-Linear Relationships

5. Differentiation

6. Higher Differentiation, Maxima and Minima

1

POWERS, INDICES AND LOGARITHMS

1.1 POSITIVE INTEGRAL INDICES

If a is any positive real number and n a positive real integer then a^n denotes the product of n factors each equal to a so that

$$a^n = a \times a \times a \times \ldots \text{ to } n \text{ factors.} \qquad (1.1)$$

For example, if $a = 10, n = 4$, then

$$a^4 = 10^4 = 10 \times 10 \times 10 \times 10 = 10,000.$$

The expression a^n in words is called the nth power of a; n is called the exponent, index or degree, and a is called the base. If $n = 2$ then a^2 is called 'a squared', and if $n = 3, a^3$ is called 'a cubed'.

Before going further it would be better to consider some of the terms that we have used in the above paragraph. We assume that all readers are familiar with the meaning of 'a positive number' and are aware of the significance of negative numbers, but we have mentioned a 'positive real number'. We deliberately use the term 'real number' because in higher mathematics this distinguishes the number from what are known as imaginary numbers. At this stage we need not worry too much about what imaginary numbers are, but briefly we may state that when a real number is squared a positive number is obtained, but if an imaginary number is squared a negative number is obtained. In other words imaginary numbers are obtained by taking the square roots of negative numbers.

Also, in the first paragraph we say that n is a positive real integer. An integer or an integral number merely denotes a whole number, i.e. a number that contains no fractions. The word 'integral' here is used as an adjective; it must not be confused with the same word used as a noun in later chapters.

For the present the indices are restricted to positive whole numbers but the base can be any positive number. The base a may have irrational values i.e. values which are non-integer and cannot be expressed as fractions, such as

π or, more often, the number denoted by e which will be considered in more detail subsequently.

Now consider the product of the two expressions a^n and a^m where a and n are as defined above and m is also a positive real integer.

By expression 1.1 we write

$$a^n = a \times a \times a \times a \ldots \text{ to } n \text{ factors,}$$

$$a^m = a \times a \times a \times a \ldots \text{ to } m \text{ factors,}$$

so that

$$a^n \times a^m = (a \times a \times a \times a \ldots \text{ to } n \text{ factors}) \times (a \times a \times a \times a \ldots \text{ to } m \text{ factors})$$

$$= a \times a \times a \times a \ldots \text{ to } n + m \text{ factors}$$

$$= a^{n+m}. \tag{1.2}$$

For example, if $a = 3, n = 5$ and $m = 2$, then

$$a^5 = 3^5 = 3 \times 3 \times 3 \times 3 \times 3 = 243$$

and

$$a^2 = 3^2 = 3 \times 3 = 9,$$

so that

$$a^m \times a^n = 3^5 \times 3^2 = (3 \times 3 \times 3 \times 3 \times 3) \times (3 \times 3)$$

$$= 3^7 = 3^{5+2}$$

$$= a^{n+m}.$$

Next consider the result of dividing a^n by a^m

$$\frac{a^n}{a^m} = \frac{a \times a \times a \times a \ldots \text{ to } n \text{ factors}}{a \times a \times a \times a \ldots \text{ to } m \text{ factors}}.$$

For the present we are considering only the case where n is greater than m so that all the m factors in the denominator cancel with m of the n factors in the numerator to give

$$\frac{a^n}{a^m} = a \times a \times a \times a \ldots \text{ to } n - m \text{ factors}$$

$$= a^{n-m}. \tag{1.3}$$

For example, if $a = 3, n = 5$ and $m = 2$, then

$$\frac{a^n}{a^m} = \frac{3^5}{3^2} = \frac{3 \times 3 \times 3 \times 3 \times 3}{3 \times 3} = 3^3$$

$$= 3^{5-2} = a^{n-m}.$$

Finally, let us consider the result of taking the mth power of the number a^n.

$$(a^n)^m = a^n \times a^n \times a^n \times a^n \times \ldots \text{ to } m \text{ factors}$$

$$= a^{n+n+n+n+\ldots \, m \text{ terms}}$$

$$= a^{nm}. \tag{1.4}$$

Again, if $a = 3$, $n = 5$ and $m = 2$ then

$$(a^n)^m = (3^5)^2 = (3.3.3.3.3) \times (3.3.3.3.3) = 3^{10}$$

$$= 3^{5 \times 2}.$$

In summary, then, if a is a positive real number and n and m are positive real integers, we have three rules for the combination of indices:

(1) Multiplication

$$a^n \times a^m = a^{n+m}$$

(2) Division

$$\frac{a^n}{a^m} = a^{n-m}$$

(3) The power of a power

$$(a^n)^m = a^{nm}.$$

Example I

Simplify

$$\frac{3^3 \times 4^3}{2^3 \times 3^4}$$

Dividing numerator and denominator by 3^3 gives

$$\frac{4^3}{2^3 \times 3}.$$

Since $4 = 2^2$ we now have

$$\frac{(2^2)^3}{2^3 \times 3} = \frac{2^6}{2^3 \times 3}$$

Dividing numerator and denominator by 2^3 we get

$$\frac{2^3}{3}$$

$$= \frac{8}{3} = 2\frac{2}{3}.$$

Example II

Simplify $\dfrac{a^3(b^2 - c^2)}{a^5 c^3 + a^5 b^3}.$

Since a^5 is a common factor of the two terms of the denominator we may write the expression in the form

$$\frac{a^3(b^2 - c^2)}{a^5(c^3 + b^3)}.$$

We divide numerator and denominator by a^3 and obtain

$$\frac{(b^2 - c^2)}{a^2(c^3 + b^3)}.$$

Numerator and denominator are now factorised to give

$$\frac{(b - c)(b + c)}{a^2(c + b)(c^2 - bc + b^2)}.$$

Finally, numerator and denominator are divided by $(b + c)$ to give

$$\frac{b - c}{a^2(c^2 - bc + b^2)}.$$

Applications

1. *Poiseuille's Formula*

The rate of flow of fluid V along a cylindrical vessel length l and radius a is given by the formula

$$V = \frac{\pi(\Delta P) a^4}{8\eta l}$$

where η is a constant characteristic of a particular fluid and known as the viscosity of the fluid, and ΔP is the difference in fluid pressure between the ends of the tube. This formula, known as *Poiseuille's formula* relates to the streamline flow of liquid along a rigid uniform cylinder. It has useful applications in the consideration of the control of the rate of blood flow along the vessels of the circulation although it is realised of course that the blood vessels are not exactly cylindrical or rigid. We can use the formula to show the changes that can be produced within those vessels that can alter their radii by way of their muscular coating. Let us consider the changes in the rate of flow when the radius a only is changed.

If we now change the radius to half its original value so that it is $\frac{1}{2}a$ the new rate of flow of fluid V' is given by

$$V' = \frac{\pi(\Delta P)(\frac{1}{2}a)^4}{8\eta l}$$

but $(\frac{1}{2}a)^4 = \dfrac{a^4}{16}$

so that

$$V' = \frac{\pi(\Delta P)a^4}{8\eta l(16)}$$

$$= V/16$$

Thus reduction of the radius by one half reduces the rate of flow by one sixteenth.

If, on the other hand, the radius is increased to $3a/2$, the rate of flow is increased by a factor $81/16$ or $5^1/_{16}$.

The formula has important consequences because of the a^4 term. As far as the circulation is concerned it means that the radius need only be changed by a small amount to produce a large change in the rate of blood flow along the vessel. A reduction in radius by just one tenth is sufficient to decrease the rate of flow to 0.6561 of its original value.

2. Growth of a Bacterial Population

Suppose that we have a single bacterium in a suitable culture medium. After a time it will divide into two daughter cells. After a further period of time, the two daughter cells will each divide to produce a total of four cells. After yet a further period of time the cells will divide to give a total of eight cells and so on. The

time interval between each successive division is known as the *generation time*.
The population doubles itself after each generation time. We may thus draw up a
table for the size of the population N at various time t expressed as multiples of
the generation time τ.

t	0	τ	2τ	3τ	4τ	5τ	$6\tau \ldots \ldots \ldots n\tau$
N	1	2	4	8	16	2^5	$2^6 \ldots \ldots \ldots 2^n$

If now instead of starting with just one bacterium the initial size of the population is large and denoted by N_0, after one generation time τ each member of
the population will have divided so that the population has doubled to $2N_0$. After
a further generation time it will be $4N_0$. After n generation times the population
will be $2^n N_0$. Obviously if we have a large population there will be no sudden
doubling up of numbers precisely at the end of each generation time but the
population will increase steadily although the mean generation time will be τ.
Provided that there is sufficient nutriment for the population to thrive, we can
get a good estimation of the size of the population $N_{n\tau}$ after n generation times
by way of the formula

$$N_{n\tau} = N_0 \, 2^n$$

As an example consider a culture which contains 1000 cells and that the generation time is 40 min. After 20 generation times the population will be

$$N_{20\tau} = 1000 \times 2^{20}$$

Now $2^{10} = 1024$, so that

$$N_{20\tau} = 1000 \times 1024 \times 1024$$

$$= 1,048,576,000$$

$$\eqsim 1000 \text{ million or } 10^9$$

where the symbol $\eqsim$ means 'nearly equal to'. . .

Thus it is seen that after 20 x 40 min, or 13 hr 20 min, the original population of
a thousand has increased by more than a million-fold.

Exercises 1.1

Simplify:

(1) $2^3 3^3 \times 2^3 3^2$

(2) $\dfrac{2^4 \, 3^3}{2^3 \, 3^2}$

(3) $\dfrac{(3^2)^3}{(3^3)^2}$

(4) $(x^2 \, y^3) \times (xy^4)$

(5) $\dfrac{(3x^3 \, y^2)^2 - (xy)^2}{3x^2 \, y + 1}$

(6) $\dfrac{4^2 \, 3^2 - 2^4}{2^3}$

(7) $\dfrac{x^3 y^4 \times 2xy}{x^7 y^2}$

(8) $\dfrac{a^3 b^4}{8c^3 d^3} \times \dfrac{(4cd^2)^3}{2ab^2}$

(9) $\dfrac{a^4(b^4 - c^4)}{(b^2 + c^2)(b - c)}$

(10) Find the value of the following function

$$\frac{a^3 b^2}{c^3} \div a^2 bc$$

(i) when $a = 4, b = 2, c = 5$,
(ii) when $a = 3, b = 2, c = 2$.

(11) A population of 1000 bacterial cells is placed for 10 hr in a culture medium in which they divide with a generation time of 1 hr. At the end of this time 0.1% of the cells are removed and placed in a different culture medium so that the generation time is 30 min. How many cells should there be in this second medium after 5 hours incubation?

(12) Use Poiseuille's formula to determine the proportional increase in the rate of blood flow through a vessel when the radius increases by one twentieth.

1.2   ZERO, FRACTIONAL AND NEGATIVE INDICES

In the previous section we considered the manipulation of quantities raised to positive integral powers and we arrived at the three rules for the combination of indices. We are now going to consider if these rules can be used to give some meaning to indices which are not positive or integral.

1. Zero indices

According to the multiplication and division rules we have

$$a^n \times a^m = a^{n+m};$$

$$\frac{a^n}{a^m} = a^{n-m}$$

If $m = 0$, and the rules still apply, we have

$$a^n \times a^0 = a^{n+0} = a^n;$$

$$\frac{a^n}{a^0} = a^{n-0} = a^n.$$

In either case it is seen that, if the equations are to hold, then $a^0 = 1$. Thus any quantity with zero index is equivalent to one, e.g. $3^0 = 1$; $10^0 = 1$.

2. *Fractional indices*

According to the multiplication rule we have

$$a^n \times a^m = a^{n+m}.$$

If this is to hold for fractional indices let us consider the case where $n = m = \frac{1}{2}$ so that

$$a^{1/2} \times a^{1/2} = a^{1/2+1/2} = a^1 = a.$$

Thus $a^{1/2}$ is a quantity which, when multiplied by itself, gives the quantity a. The quantity $a^{1/2}$ is therefore the square root of a.

$$a^{1/2} = \sqrt{a}.$$

Note. There is always a positive and negative square root to every positive number. For example, if $a = 4$ then $\sqrt{a} = \sqrt{4} = +2$ or -2.

Next let us consider the case

$$a^{1/3} \times a^{1/3} \times a^{1/3} = a^1 = a.$$

Thus $a^{1/3}$ is the cube root of a

$$a^{1/3} = \sqrt[3]{a}.$$

Next let us consider the index $1/n$ where n is a positive integer, then

$$a^{1/n} \times a^{1/n} \times a^{1/n} \ldots \text{to } n \text{ factors}$$

$$= a^{1/n+1/n+1/n \ldots \text{to } n \text{ terms}} = a^{n/n} = a.$$

Thus the interpretation of $a^{1/n}$ is the nth root of a

$$a^{1/n} = \sqrt[n]{a}$$

Example I

Evaluate $(7\frac{1}{9})^{1/2}$.

$$(7\tfrac{1}{9})^{1/2} = (\tfrac{64}{9})^{1/2} = \sqrt{\tfrac{64}{9}} = \pm\tfrac{8}{3} = \pm2\tfrac{2}{3}.$$

Example II

Evaluate $(2\tfrac{1}{4})^{3/2}$.

$$(2\tfrac{1}{4})^{3/2} = (\tfrac{9}{4})^{3/2} = (\pm\tfrac{3}{2})^3 = \pm\tfrac{27}{8} = \pm 3\tfrac{3}{8}.$$

Let us now consider the quantity $a^{n/m}$ where both n and m are positive integers. According to the law of multiplication

$$a^{1/n} \times a^{1/n} \times a^{1/n} \times a^{1/n} \times \text{to } m \text{ terms} = a^{n/m}.$$

It is thus seen that $a^{n/m}$ is the mth root of a^n.

$$a^{n/m} = \sqrt[m]{a^n}$$

If the index is for example 1.093 we can express the quantity $a^{1.093}$ in the form

$$a^{1.093} = a^1 \times a^{9/100}\, a^{3/1000} = a \times 100\sqrt{a^9} \times 1000\sqrt{a^3}.$$

Theoretically then any fractional index can be given a meaning. We shall see in later sections how logarithms can be used to calculate such quantities as $2^{4.36}$, $10^{0.12}$ etc.

3. Negative Indices

In the relationship

$$a^n \times a^m = a^{n+m}$$

let us consider the case where $m = -n$ which will give

$$a^n \times a^{-n} = a^{n-n} = a^0 = 1.$$

We may thus write

$$a^{-n} = \frac{1}{a^n}$$

For example $3^{-2} = \frac{1}{3^2} = \frac{1}{9}$; $16^{-1/4} = 1/16^{1/4} = \pm\frac{1}{2}$.

Example III

Evaluate $64^{1/2} \times 8^{-1/3} \times (\tfrac{9}{4})^{-1/2}$

Let us consider each term separately:

$$64^{1/2} = \sqrt{64} = \pm 8.$$

$$8^{-1/3} = \frac{1}{8^{1/3}} = \frac{1}{\sqrt[3]{8}} = \frac{1}{2}.$$

$$(\tfrac{9}{4})^{-1/2} = \frac{1}{(\tfrac{9}{4})^{1/2}} = \frac{1}{\sqrt{\tfrac{9}{4}}} = \frac{1}{\pm(\tfrac{3}{2})} = \pm\frac{2}{3}.$$

The product of these three terms is

$$(\pm 8)(\tfrac{1}{2})(\pm\tfrac{2}{3}) = \pm 8 \times \tfrac{1}{2} \times \tfrac{2}{3} = \pm\tfrac{8}{3}.$$

Example IV

Simplify

$$(a^3 b^6)^{2/3} \times (16a^4 b^6)^{-1/2}$$

The two terms are considered separately

$$(a^3 b^6)^{2/3} \text{ is } (\sqrt[3]{a^3 b^6})^2$$
$$= (ab^2)^2 = a^2 b^4.$$

$$(16a^4 b^6)^{-1/2} = \frac{1}{(16a^4 b^6)^{1/2}} = \frac{1}{\sqrt{16a^4 b^6}}$$

$$= \pm\frac{1}{4a^2 b^4}.$$

The product of the two terms is

$$\pm a^2 b^4 \times \frac{1}{4a^2 b^3} = \pm\frac{a^2 b^4}{4a^2 b^3}$$

Division of numerator and denominator by $a^2 b^3$ yields the final simplification to $\pm\dfrac{b}{4}$.

Exercises 1.2

Simplify:

(1) $27^{5/3}$

(2) $\dfrac{8^{1/3} \times 81^{1/4}}{36^{1/2}}$

(3) $9^{-3/2} \times 27^{2/3}$

(4) $16^{-1/2} \times 64^{1/3} \div 32^{-1/5}$

(5) $(x^6 y^4)^{1/2} \times (x^9 y^6)^{-1/3}$

(6) $(7^{0.5})^4$

(7) $64^{-1/2} + (\frac{1}{4})^{3/2} - 8^{-1}$

(8) $(4a^4 b^6)^{-3/2} \times (8a^6 b^{12})^{2/3}$.

1.3 LOGARITHMS

The logarithm of a positive real number x to a given base a is defined as the power to which the base a must be raised to equal the number x. Thus in the equation

$$a^n = x \qquad (1.5)$$

the base a is raised to the power n to equal x, so that n is the logarithm of x to the base a and is written

$$n = \log_a x$$

If the base a is positive then a^n is positive whatever the value of n. This means that we are unable to obtain the logarithm of a negative number to a positive base. For example let us consider

$$\log_2 8.$$

Now
$$8 = 2^3$$
so that

$$\log_2 8 = \log_2 (2^3)$$

and therefore by the definition of a logarithm

$$\log_2 8 = 3.$$

Likewise it may be shown that

$$\log_2 (\tfrac{1}{8}) = -3.$$

Logarithms of non-integral numbers can also be obtained. For example consider

$$\log_{10} 80.$$

Calculation of this logarithm is not usually made directly but tables are available. In fact

$$80 \simeq 10^{1.9031}$$

so that

$$\log_{10} 80 = 1.9031$$

It is not necessary that the base is a whole number either, for instance in a later section we shall consider natural logarithms which are to base e, where e is a constant having a value 2.718 (to 4 significant figures).
Thus

$$\log_e 80 = \log_{2.718} 80.$$

Tables of natural logarithms are also available but in fact

$$80 \simeq (2.718)^{4.3820} = e^{4.3820}$$

so that

$$\log_e 80 = \log_e e^{4.3820} = 4.3820.$$

Note that, if in equation 1.5 $n = 0$ then

$$a^0 = 1$$

so that by the definition of a logarithm

$$\log_a 1 = 0. \tag{1.6}$$

Thus it is seen that the logarithm of unity to any base is zero.
Also it is readily seen that for any base

$$\log_a a = 1 \tag{1.7}$$

The laws of indices are now used to obtain the rules for manipulating logarithms.

1. *Multiplication*

To prove

$$\log_a bc = \log_a b + \log_a c \tag{1.8}$$

Proof:

Put $\log_a b = x$ and $\log_a c = y$

then $b = a^x$ and $c = a^y$.

Therefore

$$bc = a^x a^y = a^{x+y}.$$

From the definition of a logarithm we may now write

$$\log_a bc = x + y$$
$$= \log_a b + \log_a c.$$

We may extend the above argument to the following result:

$$\log_a bcdef \ldots = \log_a b + \log_a c + \log_a d$$
$$+ \log_a e + \log_a f + + + +. \qquad (1.9)$$

2. Division

To prove

$$\log_a (b/c) = \log_a b - \log_a c \qquad (1.10)$$

Proof:

As before put

$$\log_a b = x \text{ and } \log_a c = y$$

then

$$b = a^x \text{ and } c = a^y$$

Therefore

$$\frac{b}{c} = \frac{a^x}{a^y} = a^{x-y}.$$

From the definition of a logarithm we may write

$$\log_a (b/c) = x - y = \log_a b - \log_a c.$$

3. The Power of a Power

To Prove

$$\log_a b^n = n \log_a b \qquad (1.11)$$

Proof:

Put

$$\log_a b = x$$

so that

$$b = a^x$$

and, therefore $$b^n = a^{nx}$$

so that $$\log_a b^n = nx = n . \log_a b.$$

4. *Change of Base*

Sometimes it is necessary to convert a logarithm to a particular base to a logarithm to another base. Suppose we know the value of $\log_a n$ and that we require to know the value of $\log_b n$. We have a rule for making the conversion

$$\log_b n . \log_a b = \log_a n \qquad (1.12)$$

Proof:

Let $$\log_b n = x$$

so that $$n = b^x.$$

Now

$$\log_a n = \log_a b^x = x \log_a b$$

$$= \log_b n . \log_a b.$$

The rule may be written in the form

$$\log_b n = \frac{\log_a n}{\log_a b}$$

Finally, consider the case where $n = a$. Equation (1.12) reduces to

$$\log_b a . \log_a b = \log_a a$$

but $\log_a a = 1$ so we obtain the result

$$\log_b a . \log_a b = 1 \qquad (1.13)$$

Example I

If $p = \log_a bc$, $q = \log_b ca$, $r = \log_c ab$,

prove that $p . q . r = p + q + r + 2$.

First of all we may write

$$p.q.r = \log_a bc . \log_b ca . \log_c ab.$$

From equation 1.8 it is seen that the right hand side of this equation may be expanded to the form

$$(\log_a b + \log_a c)(\log_b c + \log_b a)(\log_c a + \log_c b).$$

Multiplying out the brackets and remembering that $\log_a c \cdot \log_c a = 1$, we obtain

$$(\log_a b + \log_a c)(\log_b c \cdot \log_c a + \log_b a \cdot \log_c a + 1 + \log_b a \cdot \log_c b)$$

$$= \log_a b \cdot \log_b c \cdot \log_c a + \log_c a + \log_a b + \log_c b + \log_b c + \log_b a$$

$$+ \log_a c + \log_a c \cdot \log_b a \cdot \log_c b$$

The first and last terms of this expression are now simplified by using the rule for the change of base

$$\log_a b \cdot \log_b c \cdot \log_c a = \log_a b \cdot (\log_a c / \log_a b)(\log_a a / \log_a c)$$

$$= \log_a a = 1,$$

and

$$\log_a c \cdot \log_b a \cdot \log_c b = \log_a c \cdot (\log_a a / \log_a b)(\log_a b / \log_a c)$$

$$= \log_a a = 1$$

Collecting the terms together we now write

$$p.q.r. = 1 + \log_c a + \log_c b + \log_b c + \log_b a + \log_a b + \log_a c + 1$$

$$= 2 + \log_c ab + \log_b ca + \log_a bc$$

$$= 2 + p + q + r$$

Example II

Find the value of x if

$$2^{4x} - 20.2^{2x} + 64 = 0$$

If we put $Y = 2^{2x}$ the equation becomes an ordinary quadratic

$$Y^2 - 20Y + 64 = 0.$$

Upon factorisation we obtain

$$(Y - 4)(Y - 16) = 0$$

which gives the solutions

$$Y = 4 \text{ and } Y = 16$$

We consider the first of the two solutions and write

$$Y = 2^{2x} = 4 = 2^2$$

so that $$2^{2x} = 2^2$$

and therefore

$$2x = 2$$

and

$$x = 1.$$

The other solution is now considered

$$Y = 2^{2x} = 16 = 2^4$$
$$2^{2x} = 2^4$$
$$2^x = 4$$
$$x = 2$$

Exercises 1.3

(1) Prove that

$$\log_a b . \log_c a . \log_d c . \log_b d = 1$$

(2) If $p = \log_a \frac{5}{2}$, $q = \log_a \frac{10}{3}$, $r = \log_a \frac{3}{2}$ and $s = \log_a 3$, what is the value of

$$3p - 3q - 6r + 3s?$$

(3) Solve

    (a) $2^{2x} - 12.2^x + 32 = 0$

    (b) $3^{4x} - 2.3^{2x} + 1 = 0$

(4) Given that $\log_{10} 2 = 0.301$
calculate the value of $\log_{10} 5$.

(5) Solve

$$\log_a x + \log_a 10 = \log_a 2.$$

(6) Solve

$$\log_2 x = \log_8 64.$$

1.4   COMMON LOGARITHMS

At the end of this book is a table of the values of the logarithms to the base ten of numbers from 1 to 10. These tables are very useful in making rapid calculations. Logarithms to the base 10 are called *common* logarithms and instead of writing say $\log_{10} n$ we write $\log n$ and in this notation it is implied that the base is 10.

Let us consider the following numbers:

$$1, 10, 100, 1000, \ldots$$

These numbers may be written

$$10^0, 10^1, 10^2, 10^3 \ldots$$

Now by the definition of a common logarithm we may write

$$\log 1 = 0; \log 10 = 1; \log 100 = 2; \log 1000 = 3 \ldots$$

Next let us consider the following numbers:

$$0.1, 0.01, 0.001, 0.0001 \ldots$$

These numbers may be written

$$10^{-1}, 10^{-2}, 10^{-3}, 10^{-4}, \ldots$$

and so we may write

$$\log 0.1 = -1; \log 0.01 = -2; \log 0.001 = -3;$$
$$\log 0.0001 = -4 \ldots$$

Thus if we have a number which is an integral power of 10 we can immediately write down its common logarithm.

In order to determine the logarithms of non-integral powers of 10 we use the tables at the back of the book. First of all let us consider numbers between 1 and 10, and as an example consider the number 7.3. Now since we know that $\log 1 = 0$ and $\log 10 = 1$, then $\log 7.3$ is a number greater than 0 but less than 1. In the tables we go down the left hand column until we come to 73. In the column to the immediate right headed 0 is found the number 8633. This means that to four significant figures

$$\log 7.3 = 0.8633.$$

Now let us find the logarithm of 7.35. As before we go down the left hand column until we come to 73, then go along the row until we come beneath the column headed 5. The number at this point is 8663 so that

$$\log 7.35 = 0.8663$$

Finally, let us look up log 7.354. Here we repeat the process for finding log 7.35 and note its value 0.8663. Now at the right of the page of tables are columns that are called mean differences. We use the mean differences in the same row as 73 and under the mean differences column headed 4 we see the number 2. This number 2 is added to the number 8663. Thus:

$$\log 7.354 = \log 7.35 + \text{mean differences}$$
$$= 0.8663 + 0.0002$$
$$= 0.8665$$

The tables enable us to calculate logarithms to only four significant figures. If, for instance, we wanted to know the value of log 7.3548 we could not get a more accurate value than that obtained by looking up log 7.355. In other words all long numbers must be approximated to four significant figures if we are to work with four-figure logarithms.

As another example let us look up log 2.407. Going down the left hand column to row 24 and then along to the next column which is headed 0 we find the number 3802, and the mean difference under the 7 column is 12, so that

$$\log 2.407 = 0.3802 + 0.0012 = 0.3814.$$

Now let us extend the usefulness of the tables to enable us to find the logarithms of such numbers as

$$73.54, 7354, \text{ and } 7{,}354{,}000.$$

These numbers may be written in the following forms:

$$10 \times 7.354, \ 10^3 \times 7.354 \text{ and } 10^6 \times 7.354.$$

By the rule for multiplication by logarithms (Section 1.3) we may write

$$\log 73.54 = \log 10 + \log 7.354$$
$$= 1 + 0.8665$$
$$= 1.8665;$$

$$\log 7354 = \log 10^3 + \log 7.354$$
$$= 3 + 0.8665$$
$$= 3.8665;$$
$$\log 7,354,000 = \log 10^6 + \log 7.354$$
$$= 6.8665$$

The logarithm of a number is made up of two parts, an integral part called the *characteristic* and a decimal part called the *mantissa*. In each of the three numbers considered above, the mantissas of their logarithms are the same, i.e. 0.8665, but the characteristics are different. The characteristic of the logarithm of a number greater than unity is one less than the number of digits before the decimal point.

Logarithms of numbers less than unity can be found by the use of tables. Consider the numbers 0.7354, 0.007354 and 0.000007354. They may be expressed in the form

$$10^{-1} \times 7.354, \ 10^{-3} \times 7.354 \text{ and } 10^{-6} \times 7.354.$$

The logarithms of such numbers are expressed in a manner such that the mantissa is always positive. For example

$$\log 0.7354 = \log 10^{-1} + \log 7.354$$
$$= -1 + 0.8665 = -0.1335.$$

But instead of writing

$$\log 0.7354 = -0.1335,$$

we write

$$\log 0.7354 = \bar{1}.8665.$$

The symbol $\bar{1}$ (called bar-one) indicates that *only the integral part of the logarithm is negative*.

By similar reasoning we have

$$\log 0.007354 = -3 + 0.8665 + \bar{3}.8665$$
$$\log 0.000007354 = -6 + 0.8665 = \bar{6}.8665.$$

As a general rule, for numbers less than unity, the characteristic is negative and is one more than the number of zeros following the decimal point.

Exercises 1.4

Find the common logarithms of the following numbers:

(1) 7,005   (2) 7.005
(3) 700.5   (4) 0.7005

(5) 10.34   (6) 1,034
(7) 0.1034   (8) 10,349
(9) 7,941   (10) 79.44
(11) 7.9493   (12) 999
(13) 0.0001   (14) 0.0021
(15) 0.8233   (16) 0.0719

1.5   ANTILOGARITHMS

In the last section we saw how to obtain the common logarithm of any positive number. In this section we shall consider the reverse process, if we are given the logarithm of a number we want to find the number.

Now in many books of tables, tables of antilogarithms are given immediately after the tables of logarithms. These are rather unnecessary since the tables of logarithms themselves are sufficient for our purposes. For this reason no tables of antilogarithms are included at the end of this book.

Let us consider a few examples in order to show how to use the tables of logarithms in this respect.

Let us find the number whose logarithm is 0.9345. The characteristic is zero so we know immediately that the number is between 1 and 10, and from the value of the mantissa we can guess that the number is nearer 10 than 1. In the tables we go to the second page and run down the left hand column and it is seen that 0.9345 is the logarithm of 8.6.

Now let us find the number whose logarithm is given as 2.8675. The characteristic is 2 so this means that the number lies between 10^2 and 10^3, i.e. between 100 and 1000. Let us now look at the mantissa. Going down the left hand column of the logarithms we see that log 7.3 = 0.8633 and log 7.4 = 0.8692. Our required number is therefore between 730 and 740. To get the number more accurately we go horizontally along the row 73 until we see our logarithm in column 7. The number whose logarithm is 2.8675 is thus 737.

Now let us find the number whose logarithm is $\bar{3}$.2698. The characteristic is $\bar{3}$ so this means that the required number is 10^{-3} multiplied by some number which lies between 1 and 10. Looking at the tables we see that log 1.86 = 0.2695 and that log 1.87 = 0.2718. Our logarithm lies between these two values so we must go to the mean differences. Under 1 the mean difference is 2 and under 2 the mean difference is 5, so log 1.861 = 0.2695 + 0.0002 = 0.2697 and log 1.862 = 0.2695 + 0.0005 = 0.2700. Now these two values are as near as it is possible for us to get to our given mantissa of 0.2698. We must take the number whose logarithm is nearest that which we are given. Thus, as accurately as tables allow, the number whose logarithm is $\bar{3}$.2698 is 1.861×10^{-3} = 0.001861.

In finding the numbers whose logarithms are given we must remember that the mantissa is always positive. We consider the mantissa separately and find the number whose logarithm is given by it. We then look at the characteristic to find the power of 10 by which we must multiply this number.

Exercises 1.5

Find the numbers whose common logarithms are given by:

(1) 0.1761	(2) 0.1759
(3) 0.1750	(4) 0.9974
(5) 1.9983	(6) 2.0008
(7) $\bar{1}$.1397	(8) $\bar{3}$.0501
(9) 0.5100	(10) 7.9100
(11) $\bar{1}$.3302	(12) 2.0213

1.6 THE USE OF COMMON LOGARITHMS IN NUMERICAL CALCULATIONS

In order to demonstrate the usefulness of common logarithms in performing numerical calculations, certain examples are worked out. In each case before the problem is tackled a rough estimation of the answer is made so that we can detect if we have made any serious error in the proper calculation.

In making the calculation we must remember always the rules for the manipulation of logarithms, i.e. multiplication, division and power of a power. Also we must remember that, whereas the characteristic may be either positive or negative, the mantissa is always positive.

Example I

Evaluate 10.38 x 0.173 x 3.75.

A rough estimate of the answer is made first of all. 10.38 is roughly equal to 10, 0.173 is roughly 0.2 and 3.75 is roughly 4. An estimate of the answer is 10 x 0.2 x 4 = 8.

The calculation by logarithms is as follows. The logarithm of the product of three terms is the sum of the logarithms of the terms. We set out the numbers and their logarithms according to the following table:

No.	log	
10.38	1.0162	
0.173	$\bar{1}$.2380	add
3.75	0.5740	
6.733	0.8282	

10.38 x 0.173 x 3.75 = 6.73 (to 3 significant figures). We have added the three logarithms and from the tables we see that 6.733 has a logarithm which is equal to 0.8282. The answer that we obtain is of the same order of magnitude as the rough estimation. The answer is given to 3 significant figures as the fourth figure is a little unreliable.

Example II

Evaluate $\dfrac{3.692}{0.129}$

A rough estimate of the answer, as found by inspection, is 30. The numerical work is arranged in the same way as for the previous example:

No.	log	
3.692	0.5672	
		subtract
0.129	$\bar{1}$.1106	
28.61	1.4566	

$$\frac{3.692}{0.129} = 28.6 \text{ to 3 significant figures.}$$

The subtraction of the mantissas is straightforward. The subtraction of the characteristics is done according to the simple laws of algebra. What we are really saying is

$$0 - (-1) = +1.$$

Example III

Evaluate $\dfrac{0.379}{71.480}$

A rough estimate of the answer is 0.005. The arrangement of the numerical work is the same as for the previous example:

No.	log	
0.379	$\bar{1}$.5786	
		subtract
71.480	1.8542	
0.005301	$\bar{3}$.7244	

$$\frac{0.379}{71.480} = 0.00530 \text{ to 3 significant figures.}$$

In this example the subtraction of the mantissas gives a negative value. The difficulty is overcome by putting the logarithm of the number in the form

$$\bar{1}.5786 = \bar{1} + 0.5784$$
$$= \bar{2} + 1.5786$$

so that the subtraction is possible.

Example IV

Evaluate $(0.02191)^4$

One way of performing the calculation is to rewrite the expression in the form

$$(0.02191)^4 = (10^{-2} \times 2.191)^4 = 10^{-8} \times (2.191)^4$$

As a rough estimate the answer is $10^{-8} \times 16 = 1.6 \; 10^{-7}$.
As far as the logarithms are concerned we need only consider $(2.191)^4$.

No.	log	
2.191	0.3406	
		multiply
	4	
23.04	1.3624	

The required answer is thus

$$23.04 \times 10^{-8} = 2.30 \times 10^{-7} \text{ to three significant figures.}$$

Alternatively we may use logarithms directly and without prior simplification:

No.	log
0.02191	$\bar{2}.3406$ x
	4
$2.304 \; 10^{-7}$	$\bar{7}.3624$

In multiplying by 4 here we remember that the mantissa is always positive.

The mantissa and characteristic are multiplied separately and then added afterwards. So that

$$\bar{2}.3406 \times 4 = (\bar{2} \times 4) + (0.3406 \times 4)$$
$$= \bar{8} + 1.3624$$
$$= \bar{7}.3624.$$

Example V

Evaluate $(0.5318)^{1.693}$

This is roughly equal to $(0.5)^{1.5} = (0.5)^{3/2}$
$$\hat{=} (0.7)^3$$
$$\hat{=} (0.35).$$

Particular care is needed with this calculation

No.	log
0.5318	$\bar{1}.7257 \times$
	1.693

The multiplication of this logarithm by 1.693 cannot be done easily without the further use of tables. A simplification is made by multiplying the characteristic and the mantissa separately. Thus

$$1.693 \times \bar{1}.7257$$
$$= (1.693 \times \bar{1}) + (1.693 \times 0.7257)$$
$$= (-1.693) + (1.693 \times 0.7257)$$

No.	log
1.693	0.2287
0.7257	$\bar{1}.8607$
1.229	0.0894

$$= (-2 + 0.307) + (1.229)$$
$$= \bar{2} + 1.536$$
$$= \bar{1}.536$$

This last calculation gives us the result

$$\log (0.5318)^{1.693} = \bar{1}.536$$

The tables now give us the final answer

$$(0.5318)^{1.693} = 0.3436$$
$$= 0.344 \text{ to three significant figures.}$$

Another way of doing this calculation is as follows:

$$\log (0.5318)^{1.693} = 1.693 \times (\bar{1}.7257)$$
$$= 1.693 \times (-1 + 0.7257)$$
$$= 1.693 \times (-0.2743)$$

The negative sign is ignored and the product found by logarithms (a rough estimate is 0.5):

No.	log
1.693	0.2287 +
0.2743	$\bar{1}$.4383
0.4645	$\bar{1}$.6670

Thus

$$\log (0.5318)^{1.693} = -0.4645$$
$$= -1 + 0.5355$$
$$= \bar{1}.5355$$

From the tables we see

$$\log (0.3432) = \bar{1}.5355$$

and so

$$(0.5318)^{1.693} = 0.3432$$
$$= 0.343 \text{ to three significant figures.}$$

It is seen that the results of the evaluation do not quite agree after using the two methods of calculation. This arises because we are limited to four figure tables. If we had used five figure tables the answers would be in better agreement to four figures and the fifth figure in this case would show a small discrepancy. When using four figure tables we always give answers correct to three significant figures.

Exercises 1.6

(1) Use the tables of logarithms to evaluate

(a) 3.679×4.192
(b) $3.679 \times 419.2 \times 0.00981$

(c) $\dfrac{4896}{32.01}$

(d) $\dfrac{(4.623)^3}{1.2 \times 499}$

(e) $\dfrac{(5.3 \times 7.36) + 9.8}{(2.61 \times 73.982)}$

(2) In terms of the radius r the volume V of a sphere is given by the formula

$$V = \tfrac{4}{3}\pi r^3 \text{ where } \pi = \tfrac{22}{7}$$

(a) find r when V = 692 cubic inches
(b) find V when r = 2.97 mm.

(3) The surface area A of a sphere is given by the formula $A = 4\pi r^2$. If the area of a sphere is equal to 7.1 square centimetres find its radius.

(4) If one inch is equal to 2.54 centimetres, how many cubic centimetres are there in one cubic inch?

1.7   THE pH OF A SOLUTION

Pure water at ordinary temperatures dissociates very slightly into hydrogen and hydroxyl ions. The product of the concentration of hydrogen ions $[H^+]$ and the concentration of hydroxyl ions $[OH^-]$ is a constant:

$$[H^+].[OH^-] = \text{constant} = 10^{-14}. \tag{1.14}$$

Since the number of anions in the water is equal to the number of cations, we may write

$$[H^+] = [OH^-] = 10^{-7}.$$

The concentrations are expressed in g. mole per litre.

The pH of a solution is defined as the negative of the logarithm to the base 10 of the hydrogen ion concentration:

$$pH = -\log_{10} [H^+], \tag{1.15}$$

and for water

$$pH = -\log_{10} 10^{-7} = 7$$

Solutions with pH values of about 7 are called *neutral* solutions.

If an acid is added to the water, the concentration of hydrogen ions increases but relation (1.14) still holds. An increase in $[H^+]$ lowers the values of the pH according to the definition of expression (1.15). This means that the more acidic the solution, the lower is its pH value. Alternatively, if an alkali is added to the water, the concentration of hydroxyl ions is increased and this will lower the value of the hydrogen ion concentration and so raise the pH of the solution. Solutions with pH values between 7.1 and 14 are said to be *alkaline* and solutions with pH values between 0 and 6.9 are said to be *acidic*.

Example I

A biological fluid has a pH of 7.5, what is the molar concentration of hydrogen ions?
By the definition of pH we may write

$$7.5 = -\log_{10} [H^+]$$

and by the theory of logarithms we may write

$$[H^+] = 10^{-7.5} \text{ g. moles per litre.}$$

Another important relationship involving the pH of a fluid arises from the law of mass action.

If the compound CA dissociates into anions A^- and cations C^+ in solution then the relationship between the anion concentration $[A^-]$, the cation concentration $[C^+]$ and the concentration of undissociated compound $[CA]$ is given by the relationship

$$[C^+] \cdot [A^-] = K. [CA]$$

where K is a constant which is characteristic of the compound but which is temperature dependent. If we are dealing with an acid, the cation is the hydrogen ion so we may write:

$$[H^+] \cdot [A^-] = K. [HA].$$

Taking logarithms to the base 10 we may write

$$\log_{10} [H^+] + \log_{10} [A^-] = \log_{10} K + \log_{10} [HA]$$

From the definition of pH in (1.15) above we may rearrange the equation to take the form

$$pH = -\log_{10} K + \log_{10} \frac{[A^-]}{[HA]} \qquad (1.16)$$

Usually, such an equation is written in the form

$$pH = pK + \log\frac{[A^-]}{[HA]} \tag{1.17}$$

where

$$pK = -\log K \tag{1.18}$$

The value of pK is a characteristic of the reaction and its value is very important if we are considering the possible buffering properties of the compound.

For instance, the concentration of H_2CO_3 in the blood is typically 0.00125 g. mole when the pH of the solution is 7.4.
For the reaction

$$H_2CO_3 = H^+ + HC\bar{O}_3$$

the pK is 6.1 and there is virtually no dissociation of the H_2CO_3 into $H^+ + HC\bar{O}_3$.
Using equation (1.17) we may write

$$7.4 = 6.1 + \log\frac{[HC\bar{O}_3]}{[H_2CO_3]}$$

$$1.3 = \log\frac{[HCO_3^-]}{[H_2CO_3]}$$

so that

$$\frac{[HCO_3^-]}{[H_2CO_3]} = 10^{1.3} = 19.95$$

Since we are given the value of $[H_2CO_3]$ we may write

$$[HCO_3^-] = 19.95 \times 0.00125 \text{ g.mole per litre}$$
$$= 0.0249 \text{ g.mole per litre}$$

The total concentration of HCO_3^- whether in the ionic form or in combination with H^+ as H_2CO_3 is given by

$$[HCO_3^-] + [H_2CO_3] = 0.0249 + 0.02615 \text{ g. mole per litre} \tag{1.19}$$

The total concentration of hydrogen ion whether in the form of free hydrogen ion or in combination with HCO_3^- to form H_2CO_3 is given by

$$[H^+] + [H_2CO_3] = 10^{-7.4} + 0.00125 \text{ g. mole per litre}$$

$$\simeq 0.00125 \text{ g. mole per litre}$$

In this case $[H^+]$ is much smaller than $[H_2CO_3]$ and is neglected.

Consider now what would happen if we add a strong acid, i.e. one which completely dissociates such as H.Cl, so that the pH is reduced to 7.1. We denote the concentrations of the components of the solution at this new pH by dashes, e.g. $[H]'$.

By equation (1.17) we write

$$7.1 = 6.1 + \log \frac{[HCO_3^-]'}{[H_2CO_3]'}$$

From this we obtain

$$\frac{[HCO_3^-]'}{[H_2CO_3]'} = 10. \tag{1.20}$$

The sum of $[HCO_3^-] + [H_2CO_3]$ has not changed, so we may use equation (1.19) to write

$$[HCO_3^-] + [H_2CO_3] = [HCO_3^-]' + [H_2CO_3]'$$
$$= 0.02615 \text{ g. mole per litre.} \tag{1.21}$$

Equations (1.20) and (1.21) are now combined to give

$$10[H_2CO_3]' + [H_2CO_3]' = 0.02615$$

or

$$[H_2CO_3]' = 0.00238$$

The total hydrogen ion in the solution whether in the free ionic form or in combination with HCO_3^- as H_2CO_3 is given by

$$[H^+]' + [H_2CO_3]' = 10^{-7.1} + 0.00238$$
$$= 0.00238$$

The amount of hydrogen ion that was added to reduce the pH from 7.4 to 7.1 is equal to

$$0.00238 - 0.00125 = 0.00113 \text{ g.mole per litre.}$$

Since the acid fully dissociates, this represents the addition of 0.00113 g. mole of H.Cl. If this amount of acid had been added to a litre of solution at pH 7.4 which contains no H_2CO_3 the pH would be changed to

$$-\log_{10} 0.00113 = -(\bar{3}.0531)$$
$$= -(-2.9469)$$
$$2.95.$$

This example serves to illustrate how, by the presence of a buffer such as H_2CO_3 the pH changes in a biological fluid can be maintained at about 7.1 – 7.4. The amount of pH change when acid is added or removed depends upon the concentration of buffer present.

Exercises 1.7

(1) Calculate the values of the pH for the solutions which contain the following concentrations of hydrogen ions (expressed in g.moles per litre):

(a) 0.00456
(b) 0.0000062
(c) 0.017
(d) 0.000000000719
(e) 0.0000031
(f) 0.000029

(2) The pH values of certain solutions are given below. Calculate the hydrogen ion concentration in g.moles per litre:

(a) 5.3   (b) 11.8   (c) 7.0   (d) 2.9   (e) 6.9   (f) 5.8   (g) 10.1
(h) 6.2.

(3) The pH of a solution is changed from 11.8 to 11.6. Calculate the change in hydrogen ion concentration.

(4) With the usual notation the formula for the pH of a solution is given by

$$pH = pK + \log \frac{[A^-]}{[HA]}$$

Use this equation to find the values of the appropriate quantities in order to complete the following table.

	pH	pK	[A⁻]/[HA]
(a)	7.1		5
(b)		6.12	10
(c)	7.4	6.12	
(d)		6.12	7
(e)	7.1	6.20	
(f)	6.9	6.20	
(g)		6.00	20
(h)	7.05	6.15	
(i)	7.08		30

(5) A certain solution contains a buffer which has a pK value 5.2 and the concentration of undissociated buffer at pH 6.1 is 0.00001 g.mole per litre.

Calculate the concentration of strong acid which must be added to reduce the pH to 6.0. How much hydrogen ion per litre must be removed to increase the pH from 6.1 to 6.3? Suppose that the *total* amount of buffer is reduced to 0.000001 g.mole per litre, calculate how much strong acid is needed to reduce the pH from 6.1 to 6.0.

1.8   THE BODY SURFACE AREA

The body surface area of an animal is a quantity which has physiological significance from the point of view of studies involving heat loss. It is a very tedious process to take actual measurements from an animal to determine its surface area especially if several animals are involved in the experiment. The difficulty has been largely overcome by devising an empirical formula in which the surface area is related to two more easily measurable quantities. In the case of the human, an empirical formula for the surface area was devised by DuBois and DuBois (1916):

$$S = 0.007184(W)^{0.425}.(H)^{0.725} \qquad (1.22)$$

where S is the body surface area measured in square metres,
   W is the weight of the body in kilograms, and
   H is the height in centimetres.
In order to be able to use such a formula logarithms are taken of both sides.

$$\log S = \log 0.007184 + 0.425 \log W + 0.725 \log H \qquad (1.23)$$

Example 1

Calculate the body surface area of a man of height 2 metres and weight 70 kilograms.
The appropriate values are inserted into expression (1.23) to give

$$\log S = \log 0.007184 + 0.425 \log 70 + 0.725 \log 200$$

The terms on the right hand side of the equation are considered separately

   (a) $\log 0.007184 = \bar{3}.8563$
   (b) $\log 70 = 1.8451$

We use logarithms to calculate 0.425 log 70.

No.	log
0.425	$\bar{1}.6284$
1.8451	0.2660
0.7841	$\bar{1}.8944$

Thus 0.425 log 70 = 0.7841

    (c) log 200 = 2.3010

We use logarithms to calculate 0.725 log 200

No.	log
0.725	$\bar{1}.8603$
2.3010	0.3619
1.6680	0.2222

Thus 0.725 log 200 = 1.6680

The three quantities are now added together to give

$$\log S = \bar{3}.8563 +$$
$$0.7841$$
$$1.6680$$
$$\overline{0.3084}$$
$$S = 2.03 \text{ sq. metres.}$$

Exercises 1.8

(1) Determine the surface area for the following pairs of values of W and H by using the formula 1.22

	H cms	W Kg
(a)	200	70
(b)	200	75
(c)	180	70
(d)	180	65
(e)	170	60

(2) If the surface area is measured in square inches, W measured in pounds and H in inches, what is the new value of the constant term in equation (1.22)? (1 inch $\triangleq$ 2.54 cm; 1 lb $\triangleq$ 453 gm.)

1.9   BEER'S LAW

When light is allowed to pass through a solution, the emerging light is of lower intensity than the incident light because some of the light is absorbed by the solution. The relationship between the incident light of intensity I_0 and the emerging light of intensity I is given by Beer's Law

$$I = I_0 \, 10^{-\epsilon c d} \qquad (1.24)$$

where c is the concentration of the solution in moles per litre, d is the thickness of the liquid through which the light passes and ϵ is called the extinction coefficient for the solute for light of a particular wavelength.

This law has a useful application in biochemical analysis because it enables us to measure concentrations of a solute simply by making measurements of the intensity of a suitable beam of light that has passed through the fluid. An instrument for this purpose is called a *colorimeter*. It consists essentially of a light source which is made monochromatic by the use of an appropriate filter and the light then passes through a suitable cell or cuvette which contains the test solution. The intensity of the beam after passing through the fluid is recorded by means of a photoelectric cell, the output from which is recorded on a galvanometer.

Initially the cuvette is filled with distilled water and the intensity of the light is so adjusted that it gives a full-scale galvanometer reading. A standard solution of a known concentration is next inserted and the reading noted.

Finally the test solution is inserted and the reading noted.

For the standard solution we have

$$I_{st} = I_0 . 10^{-\epsilon c_{st} d} \qquad (1.25)$$

where I_{st} is the intensity of the emerging light after passing through the standard solution of known concentration c_{st}. For the test solution we have

$$I_T = I_0 . 10^{-\epsilon c_T d} \qquad (1.26)$$

where I_T is the intensity of the emerging light after passing through the test solution of unknown concentration c_T.

By taking logarithms to the base 10 of expressions (1.25) and (1.26) we obtain

$$c_{st} = \frac{\log I_0 - \log I_{st}}{\epsilon d}$$

and

$$c_T = \frac{\log I_0 - \log I_T}{\epsilon d}$$

We can eliminate ϵd and obtain

$$\frac{c_T}{c_{st}} = \frac{\log I_0 - \log I_T}{\log I_0 - \log I_{st}} \tag{1.27}$$

The scale on the instrument indicates the current generated by the photocell and the deflection of the needle is proportional to the light transmitted by the solution. Suppose that the instrument is adjusted to give full scale deflection, i.e. to read 100, when the distilled water is inserted. The reading for any other solution will then be a percentage of the incident light that passes through it.

Equation (1.27) can now be rewritten

$$\frac{c_T}{c_{st}} = \frac{\log(100/I_T)}{\log(100/I_{st})} = \frac{R_T}{R_{st}} \tag{1.28}$$

where $R_T = \log(100/I_T)$ and $R_{st} = \log(100/I_{st})$.

Now most instruments have two sets of calibrations on the scale. The second calibration gives a value of $\log(100/I)$ so that there is no need to calculate the logarithms involved.

From expression (1.28) we obtain

$$c_T = \frac{R_T \cdot c_{st}}{R_{st}}. \tag{1.29}$$

By the use of this formula we can calculate the concentration of the test solution since c_{st} is known and the two values of R are read off from the instrument.

Exercises 1.9

Light of a particular wavelength was passed through a cell 5.0 cm thick containing 0.01 molar solution of a given substance; it was found that the intensity of the transmitted light was 0.245 of the intensity of the incident light.

(a) Calculate the extinction coefficient,
(b) How thick should the cell be to produce a 50% reduction in intensity?
(c) What would be the reduction in intensity through a cell 5.0 cm thick if the concentration was 0.1 molar solution?

2

EXPONENTIAL SERIES

2.1 FACTORIALS

Consider the series of numbers

$$1,2,3,4,5,6.$$

This sequence is known as the first six natural numbers. If we multiply them all together we obtain

$$1 \times 2 \times 3 \times 4 \times 5 \times 6 = 720.$$

As a form of shorthand we may write

$$6! = 1 \times 2 \times 3 \times 4 \times 5 \times 6$$

where 6! is called *'factorial 6'*. In some books factorial 6 may be written $\lfloor 6$ but this is not very common nowadays.

Consider now the sequence

$$1,2,3,4.$$

These are the first four natural numbers and we may write

$$4! = 1 \times 2 \times 3 \times 4 = 24.$$

Similarly we may consider any sequence of the first n natural numbers

$$1,2,3, \ldots (n-2), (n-1), (n)$$

and write

$$n! = 1 \times 2 \times 3 \ldots (n-2) . (n-1) . (n)$$

where n is a positive integer. Strictly speaking n need not necessarily be a positive integer but such other cases will not be considered in this book.

Now let us consider the two factorials

$$6! = 1 \times 2 \times 3 \times 4 \times 5 \times 6$$

and

$$5! = 1 \times 2 \times 3 \times 4 \times 5.$$

We can combine them in the following way:

$$6! = 6 \times 5!.$$

If now, instead of 6! and 5! we have $n!$ and $(n-1)!$ we may write

$$n! = n \cdot (n-1)! \qquad (2.1)$$

This is known as a *'reduction formula'* and it may be extended as follows:

$$n! = n \cdot (n-1) \cdot (n-2)!$$

Division of one factorial by another may be carried out as for example

$$\frac{7!}{4!} = \frac{7 \times 6 \times 5 \times 4!}{4!} = 7 \cdot 6 \cdot 5 = 210.$$

Let us now consider the general case and divide $n!$ by $r!$ where r is an integer less than n:

$$\frac{n!}{r!} = \frac{1 \times 2 \times 3 \ldots (r-2).(r-1).r.(r+1) \ldots (n-1).n}{1 \times 2 \times 3 \ldots (r-2).(r-1).r}$$
$$= (r+1).(r+2) \ldots (n-2).(n-1).n.$$

Example I

Expand $\dfrac{(4n)!}{(2n)!}$.

This is initially expanded in the form

$$\frac{4n.(4n-1).(4n-2).(4n-3).(4n-4) \ldots}{2n.(2n-1).(2n-2).(2n-3).(2n-4) \ldots}$$

Taking first, fifth, ninth etc. terms in the numerator together we have

$$4n \cdot (4n-4) \cdot (4n-8) \ldots$$
$$=$$
$$4n \cdot 4 \cdot (n-1) \cdot 4 \cdot (n-2) \cdot 4 \cdot (n-3) \ldots = 4^n \cdot n!$$

Taking third, seventh, eleventh, etc. terms in the numerator together we have

$$(4n - 2) (4n - 6) (4n - 10) \ldots$$

$$= 2^n . (2n - 1) (2n - 3) (2n - 5) \ldots$$

Taking first, third, fifth, etc. brackets in the denominator together we have

$$2n . (2n - 2) (2n - 4) \ldots$$

$$= 2^n . n . (n - 1) (n - 2) \ldots = 2^n . n!$$

Therefore the expression becomes

$$\frac{4^n . n! \, 2^n (2n - 1)(2n - 3) \ldots (4n - 1)(4n - 3) \ldots}{2^n . n! (2n - 1)(2n - 3) \ldots}$$

$$= 4^n . (4n - 1)(4n - 3)(4n - 5) \ldots$$

Exercises 2.1

(1) Evaluate

(a) $\dfrac{8!}{4! \, 4!}$

(b) $\dfrac{(6! - 5!)^2}{7! \, 4!}$

(c) $\dfrac{1}{4!} - \dfrac{1}{5!}$

(d) $\dfrac{1}{3!} + \dfrac{1}{4!}$

(e) $6! \left(\dfrac{1}{5!} - \dfrac{1}{6!} \right)$

(2) Evaluate

(a) $\dfrac{1}{2!} + \dfrac{1}{4!} + \dfrac{1}{6!}$

(b) $\dfrac{1}{2!} + \dfrac{1}{3!} - \dfrac{1}{4!}$

(c) $\dfrac{n!(n + 1)!}{(n - 3)!(n + 2)!}$

(3) Express in terms of factorials

(a) $n . (n - 1) (n - 2) (n - 3) \ldots (n - r + 1)$.

(b) $n . (n + 1) (n + 2) (n + 3) \ldots (n + r)$

(4) Show that

(a) $1 \times 3 \times 5 \times 7 \ldots (2n-1) = \dfrac{(2n)!}{2^n . n!}$

(b) $\dfrac{11}{2} \times \dfrac{9}{2} \times \dfrac{7}{2} \times \dfrac{5}{2} \times \dfrac{3}{2} \times \dfrac{1}{2} = \dfrac{12!}{2^{12} . 6!}$

2.2. THE EXPONENTIAL SERIES

The exponential function e^x or exp $[x]$ is one of the most important functions in higher mathematics and it occurs very often in work connected with biological problems. The exponential function can be expressed in the following form:

$$e^x = 1 + x + \frac{x^2}{2!} + \frac{x^3}{3!} + \frac{x^4}{4!} + + + \qquad (2.2)$$

This expression is an infinite series—it contains an infinite number of terms, and x can have any value but it must be a *dimensionless* quantity.

If we choose $x = 1$, we define the number e and, from the series above we see that

$$e = 1 + 1 + \frac{1}{2!} + \frac{1}{3!} + \frac{1}{4!} + + + + + \qquad (2.3)$$

$$= 1 + 1 + \frac{1}{2} + \frac{1}{3 \times 2} + \frac{1}{4 \times 3 \times 2} + + +$$

$$= 1 + 1 + \frac{1}{2} + \frac{1}{6} + \frac{1}{24} + + + +$$

$$= 2.718 \text{ correct to 4 significant figures.}$$

This number, like the number π, cannot be exactly expressed. We can only give an approximation however many figures we like to choose. If we extended the accuracy to eight figures we would find

$$e = 2.7182818$$

Thus the number e like π or $\sqrt{2}$, belongs to the class known as irrational numbers—it cannot be expressed in the form m/n where both m and n are integers.

Let us now consider why such a function is called an exponential function in view of what we already know about exponents. At the same time we will justify the way it is written i.e., 'e raised to the power of x'. First of all, since we have said that x may be any number, we may choose another number y so that we can write

$$e^y = 1 + y + \frac{y^2}{2!} + \frac{y^3}{3!} + \frac{y^4}{4!} + + + +$$

$$e^x = 1 + x + \frac{x^2}{2!} + \frac{x^3}{3!} + \frac{x^4}{4!} + + + +$$

If we carefully multiply these two series together and then group all terms of the same power we have

$$e^y.e^x = 1 + (x+y) + \left(\frac{x^2}{2!} + \frac{y^2}{2!} + xy\right)$$

$$+\left(\frac{x^3}{3!} + \frac{x^2 y}{2!} + \frac{xy^2}{2!} + \frac{y^3}{3!}\right)$$

$$+\left(\frac{x^4}{4!} + \frac{x^3 y}{3!} + \frac{x^2 y^2}{2!2!} + \frac{xy^3}{3!} + \frac{y^4}{4!}\right) + + +$$

$$= 1 + (x+y) + \frac{(x+y)^2}{2!} + \frac{(x+y)^3}{3!} + \frac{(x+y)^4}{4!} + +$$

From the definition of the exponential series we may write

$$e^{(x+y)} = 1 + (x+y) + \frac{(x+y)^2}{2!} + \frac{(x+y)^3}{3!} + \frac{(x+y)^4}{4!} + +$$

so that

$$e^x.e^y = e^{(x+y)} \tag{2.4}$$

It is thus noted that the exponential function obeys the *index* or *exponent* law for products and for this reason it is called the exponential function.

In equation (2.4) it was assumed that x and y could be any real numbers and accordingly they may take negative values. Let us then put $y = -x$, where x is positive, in equation (2.4) to give

$$e^x.e^{-x} = e^{(x-x)} = e^0 = 1$$

$$\text{so that } e^{-x} = \frac{1}{e^x} \tag{2.5}$$

Also we may put $y = -z$, where z is positive, in equation (2.4) and obtain

$$e^x.e^{-z} = e^{(x-z)}$$

but according to expression (2.5) we may write:

$$e^x . e^{-y} = \frac{e^x}{e^y} = e^{x-y} \qquad (2.6)$$

In fact it can be shown that the rules for the combination of indices apply in the case of the exponential function viz:

Multiplication $e^x . e^y = e^{x+y}$;

Division $\dfrac{e^x}{e^y} = e^{x-y}$;

Power of a power $(e^x)^y = e^{xy}$. $\qquad (2.7)$

and we note at the same time that $e^0 = 1$.

The values of e^n for various values of n are given in tables at the end of this book.

Exercises 2.2

(1) By the use of the exponential series of expression (2.2) calculate the values of the following quantities to four significant figures

   (a) $e^{1/2}$

   (b) $e^{0.1}$

   (c) e^{-2}

(2) Given that $e^3 = 20.09$ and $e = 2.718$ calculate the values of (a) e^4, (b) e^{-2}, (c) e^6, (d) e^5. Give the answers to three significant figures

2.3 NATURAL LOGARITHMS

Owing to the importance of the number e in higher mathematics, logarithms to the base e have been found to be useful also. Logarithms of this type are called 'natural logarithms', 'hyperbolic logarithms' or, sometimes 'Naperian logarithms' after Lord Napier their inventor. Mathematicians often use a special notation for natural logarithms—instead of writing $\log_e N$, the shortened form $\ln N$ is used.

Natural logarithms are related to common logarithms by the formula for the change of base which was discussed in the first chapter

$$\log N = \frac{\log_e N}{\log_e 10} = \frac{\ln N}{\ln 10} \qquad (2.8)$$

Since the value of ln 10 is 2.3026 equation (2.8) takes the form

$$\ln N = 2.3026 \log N \qquad (2.9)$$

In words we may write 'to convert common logarithms to natural logarithms multiply the common logarithm by 2.3026'.

Tables are given at the end of this book for the natural logarithms of numbers between 1 and 10. These tables are used in a rather different way from the tables of common logarithms. In the first place the characteristic of the logarithms is also given but this is indicated only in the first column of figures. For instance, the natural logarithm of 3.9 is given in the tables as 1.3610, but the logarithm of 3.91 appears as .3635. In the latter case we have to remember that the logarithm has the characteristic which is given for 3.9 so that

$$\ln 3.91 = 1.3635.$$

Care in deciding the characteristic must be taken when looking up the natural logarithms of numbers between 2.7 and 2.8. Reference to the tables will show that the logarithm of 2.71 is 0.9969, for 2.72 it is 1.006 and for 2.73 it is given as .0043. In this last case we must add 1 to the logarithm to give ln 2.73 = 1.0043. All other logarithms of numbers between 2.74 and 2.79 have to have 1 added to the logarithm given in the tables. Similar conditions apply when we look up the logarithms of the numbers between 7.38 and 7.39 when the characteristic of the logarithms changes from 1 to 2.

Apart from these points the natural logarithms of numbers between 1 and 10 are obtained in much the same way as the common logarithms of these numbers.

Let us now consider how we can use the tables to find the natural logarithms of numbers greater than 10.

Take for instance the number 7398.

$$\ln 7398 = \ln (7.398 \times 10^3)$$

$$= \ln 7.398 + \ln 10^3$$

From the tables we see that

$$\ln 7.398 = 2.0012$$

Also, at the foot of the tables we are given the natural logarithms of various powers of 10 and in particular we see

$$\ln 10^3 = 6.9078$$

Thus

$$\ln 7398 = 2.0012 + 6.9078$$
$$= 8.9090.$$

In other words, in order to find the natural logarithms of numbers greater than 10 we rearrange the number in the form of a number between 1 and 10 which is multiplied by a power of 10.

The logarithms of numbers less than 1 are dealt with according to the following example:

$$\ln 0.007398 = \ln (7.398 \times 10^{-3})$$

$$= \ln 7.398 - \ln 10^3$$

$$= 2.0012 - 6.9078$$

$$= -4.9066.$$

An important point to note here is that both the characteristic and the mantissa may be negative. The characteristic is not expressed with a bar over it.

In other words, in order to find the natural logarithm of a number less than unity, first express the number as a number between 1 and 10 divided by a power of 10.

Sometimes we are given the natural logarithm and we need to find the number that it represents. The method of doing this is explained by way of the following examples.

Let N be the number whose natural logarithm is 4.8230. Then

$$\ln N = 4.8230.$$

If we look at the logarithms of powers of 10 at the foot of the tables we see that $\ln N$ is greater than $\ln 10^2$ but less than $\ln 10^3$. Since

$$\ln 10^2 = 4.6052$$

we may write

$$\ln N = 4.6052 + 0.2178$$

We now use the tables to find the number whose natural logarithm is 0.2178. On looking at the tables we see that $\ln 1.24 = 0.2151$ and $\ln 1.25 = 0.2231$. By reference to the mean difference tables we further see that to the nearest approximation the number 1.243 has a natural logarithm value 0.2178. We are now in a position to write

$$\ln N = \ln 10^2 + \ln 1.243$$

$$= \ln 124.3$$

Therefore $N = 124.3.$

Let us now consider a number whose natural logarithm is negative. To find the number whose natural logarithm is -7.9470.

Here we must remember that the values of the logarithms in the main part of our tables are positive and we must alter our logarithm so that it contains both a positive part and a negative part. If we look to the foot of the tables we see that the required number must lie between 10^{-3} and 10^{-4}. We take the value of ln 10^{-4} and write:

$$\ln N = -7.9470$$

$$= -7.9470 - \ln 10^4 + 9.2103$$

$$= 1.2633 - \ln 10^4$$

$$= \ln 3.537 - \ln 10^4.$$

Therefore

$$N = 3.537 \times 10^{-4}$$

$$= 0.0003537.$$

Exercises 2.3

(1) Find the natural logarithms of the following numbers:

(a) 79.3	(b) 0.4971	(c) 0.0044	(d) 1297
(e) 2.713	(f) 72.81	(g) 890100	(h) 0.0111
(i) 42.99	(j) 1.314	(k) 3.142	(l) 10^{-5}
(m) 112.1			

(2) Find the numbers whose natural logarithms have the following values:

(a) 3.1000	(b) -4.2000	(c) -0.4230	(d) 1.2090
(e) 17.7791	(f) 0.2311	(g) 4.8991	(h) 0.7713

2.4   EXPONENTIAL GROWTH AND DECAY

In the early sections of the first chapter we considered such functions as a^n where a is any positive real number and n is any positive or negative number. Let us now, in addition, consider the number k such that

$$k = \ln a.$$

This expression may be rearranged to take the form

$$a = e^k$$

We may now write

$$a^n = (e^k)^n = e^{nk} \tag{2.10}$$

Thus, by the use of the above expressions, we can convert any function of the type a^n to the form of an exponential.

On many occasions the biologist meets equations concerned with for instance population growth or the decay of radioactive matter which are expressed in the form

$$y = A\,e^{kt} \text{ or } A\exp[kt] \tag{2.11}$$

where A and k are constants and t represents time. In all such equations which have exponents it must be realised that the exponents are always dimensionless quantities and, since the exponent in expression (2.11) is kt, the constant k must have the dimension of reciprocal time. Often expression (2.11) is written in the form

$$y = A\,e^{t/\tau} \tag{2.12}$$

where τ has the dimension of time and it is called the time constant of the exponential rise or decay.

If, in expression (2.12), $t = 0$ then

$$y = A$$

and if $t = \tau$ then

$$y = A e.$$

The time constant τ represents the time taken for the function to become e times its original value.

If, on the other hand, we have the function

$$y = A\,e^{-t/\tau} \tag{2.13}$$

Then at $t = 0$ we have

$$y = A$$

and at $t = \tau$

$$y = A\,e^{-1} = \frac{A}{e}.$$

In this case the time constant represents the time taken for the function to decrease to $\frac{1}{e}$ of its original value.

Quite often we need to compare functions of the type

$$a^{kt}, \, b^{nt}, \, c^{mt} \text{ etc.}$$

If these functions are retained in the above form the comparison is difficult, but suppose we express each of these functions as exponentials by using expression (2.10). The functions can then be written

$$\exp(t/\tau_1), \, \exp(t/\tau_2), \exp(t/\tau_3) \text{ etc.}$$

We would now be in a better position to make comparisons between the functions. For instance, we can say that the first increases by e times its present value after a time τ_1, the second after a time τ_2 and the third after a time τ_3 etc.

For example, consider the functions

$$3^{3t}, \, 2^{4t}.$$

For the first of these we put

$$k = \ln 3 = 1.0986$$

so that

$$3 = e^k = e^{1.0986}$$

and

$$3^{3t} = \exp[(1.0986).3t] = \exp[3.2958.t]$$
$$= \exp[t/0.3034]$$

We next consider the function 2^{4t}.

Now $$\ln 2 = 0.6931$$

so that

$$2 = \exp[0.6931]$$

and

$$2^{4t} = \exp[2.7724t] = \exp[t/0.3607].$$

In their original form it is difficult to compare the rates of growth of the functions. However, when they are expressed in exponential form it is seen that the respective time constants are 0.3034 and 0.3607. If the time constant is measured in seconds this means that the former function increases to e times its original value after 0.3034 seconds but the second function takes 0.3607 seconds to increase to e times its original value.

The time constants are thus used to compare the rates of growth or decay of different functions of the form e^t/τ.

Exercises 2.4

(1) A bacterial population is increasing exponentially so that at a time t the population has a size $A \exp[t/\tau]$. A second population has a size $\frac{1}{2}A \exp[2t/\tau]$ at time t. Which population has attained the greater size at times $t = 0.1\tau$, 0.5τ and τ? At what time will the populations be of equal size?

2.5   RADIOACTIVE DECAY

Radioactive substances are used extensively in biochemical studies for investigating the various chemical reactions that take place in living organisms. A radioactive substance is one whose atoms emit either alpha-rays, which have a positive charge, and which are in fact doubly ionised helium nuclei, or beta-rays, which are swiftly moving electrons, or gamma-rays which are similar in nature to X-rays. Various methods have been devised for measuring the amount of radiation emitted by an active substance. A substance which emits radiation is said to be undergoing radioactive decay. If the amount of radioactive substance present in a sample is measured over a period of time it is found that it decreases exponentially. Thus if C is the amount of substance emitting radiation at a particular time and if C_0 is the amount of substance at a zero point in time, then the following relationship holds:

$$C = C_0 \exp[-kt] \qquad (2.14)$$

where k is a constant and depends upon the nature of the radioactive material. It is seen from this relationship that, although some substances might decay more rapidly than others, there is no theoretical point in time at which all the substance has decayed.

Radioactive substances are characterised by a 'half-life'; it is the time taken for a quantity of radioactive material to be reduced by one half.

Thus in equation (2.14) put $C = C_0/2$, so that the time taken for C_0 to be reduced to half its value is $t_{1/2}$ and

$$\frac{C_0}{2} = C_0 \exp[-k.t_{1/2}]$$

so that

$$\tfrac{1}{2} = \exp[-k.t_{1/2}]$$

and

$$\ln\tfrac{1}{2} = -k \cdot t_{1/2}.$$

Therefore

$$t_{1/2} = -\frac{1}{k}\ln\tfrac{1}{2} = \frac{0.693}{k} \tag{2.15}$$

The half life is related to k and if we know the half life, we can calculate k and thus specify the whole time course of the decay.

A radioactive substance decays to one half its amount after one half-life, to one quarter its amount after two half-lives, and to one eighth of its amount after three half-lives and so on.

The values of the half-lives of the various radioactive substances lie in the range 10^{-9} seconds to over 10^{9} years! (A selection of half-lives for various radioisotopes is given on p. 285).Those substances with a half-life of extremely short duration are of no use to the biologist because the substance decays so rapidly to non-measurable levels.

Potassium-42 has a half-life of 12.45 hr so if the experiment is of long duration a correction will have to be made for the decay in order to distinguish the reduction in amount from any reduction which genuinely arises due to the nature of the experiment.

Example I

A nerve cell is injected with a quantity of Potassium-42. After 1 hr the amount of Potassium-42 is reduced to 89.5% of its original value. Has any of the radioactive material been removed from the cell?
The half-life is 12.45 hr so that from expression (2.15) above we obtain the value for k

$$k = \frac{0.693}{t} = \frac{0.693}{12.45}$$

From equation (2.14) above we obtain the amount of substance remaining after 1 hour if all of it remains in the cell:

$$C = C_0 \exp\left[-\frac{0.693}{12.45} \times 1\right]$$

$$\ln\frac{C}{C_0} = -\frac{0.693}{12.45} \simeq -0.0557$$

From the tables we obtain

$$\frac{C}{C_0} = 0.945$$

Thus, if the loss were due to radioactive decay alone, 94.5% of the substance would still be in the cell after 1 hour. The fact that only 89.5% remains in the cell suggests that the cell has lost or removed some of the injected potassium.

A radioactive substance of importance in studies in the metabolism of carbohydrate is C^{14} which has a half-life of 5760 years. It is never necessary to take into account the amount of substance which has undergone decay when assessing the results of an experiment for in such cases the amount of activity remaining after a period as long as one year is not significantly different from that in the sample initially.

Example II

Suppose a radioactive source contains 1 million radioactive atoms. The half-life is 5760 years. How many of these atoms will be expected to have decayed after one year?
From equation (2.15) we write

$$k = \frac{0.693}{5670}$$

From expression (2.14) we write

$$C = C_0 \exp\left[-\frac{0.693}{5760} \cdot 1\right].$$

The amount which would be expected to decay after 1 year is given by $(C_0 - C)$.

$$C_0 - C = C_0\left(1 - \exp\left[\frac{0.693}{5760}\right]\right)$$

$$\simeq 10^6\left(1 - \left[1 - \frac{0.693}{5760}\right]\right)$$

(because third and higher terms of the exponential expansion are insignificant)

$$= 10^6 \cdot \frac{0.693}{5760}$$

$$= \frac{6.930}{5.760} \cdot 10^2$$

$$\simeq 120$$

Thus, on average, one atom decays about every three days if we have an initial sample of 1 million atoms.

Exercises 2.5

(1) Phosphorous-32 has a half-life of 14.2 days. How long would it take a sample to lose (a) 1% of its activity, (b) 90% of its activity?

(2) A biochemical preparation is incubated with ATP whose terminal phosphate group contains phosphorous-32 (half-life 14.2 days). After 24 hours incubation free radioactive phosphate is found to be present. What correction has to be applied in order to get an accurate estimate of the amount of ATP that has broken down?

2.6 FREE ENERGY CHANGES OF REACTIONS

The higher animals maintain themselves at a temperature of about $37°C$. and therefore in their chemical reactions they cannot rely on an external source of heat as a steam engine does. It is possible however, for living cells to obtain energy from chemical reactions in order to perform work under constant conditions of temperature and pressure. The chemical work involved is determined by what are known as free energy changes. Free energy is denoted by the symbol F (or, sometimes, G).

Suppose that we have two substances A and B which react to form two products X and Y. In such a reaction there is usually a change in free energy as the reactants change to the products. For instance when carbon is burnt in oxygen to form CO_2 an amount of energy is given out as heat. The reaction produces about 100 k.cals of energy per mole. Since, in this case, heat is given out by the reaction the change in free energy of the system denoted by ΔF is negative and the reaction is said to be *exergonic*. On the other hand if we want to break down CO_2 to form carbon and oxygen, energy must be added to the system and in this case ΔF is positive, and the reaction is said to be *endergonic*.

Consider the reversible reaction

$$A + B \rightleftharpoons X + Y \qquad (2.16)$$

then at the equilibrium concentrations, the net rates of formation of X and Y will be zero and the concentrations of the products and the reactants define the equilibrium constant K of the reaction:

$$K = \frac{[X]_e [Y]_e}{[A]_e [B]_e} \qquad (2.17)$$

where $[\ ]_e$ denotes the molar concentration at the equilibrium condition.

In reaction (2.16) it is assumed that one molecule of each substance is involved. If instead we have

$$aA + bB + + + \rightleftharpoons xX + yY + + +$$ (2.18)

where $a, b, \ldots x, y, \ldots$ represent the numbers of molecules of each substance involved in the reaction, then we write

$$K = \frac{[X]_e^x [Y]_e^y \ldots}{[A]_e^a [B]_e^b \ldots}$$ (2.19)

Suppose that we start off with reaction (2.18) but that it is not in the equilibrium state. The reaction can go to equilibrium but the system will undergo a change in free energy ΔF. The formula for the change in free energy is

$$\Delta F = RT \ln \left\{ \frac{[X]^x [Y]^y \ldots}{[A]^a [B]^b \ldots} \right\} - RT \ln K$$ (2.20)

where R is the gas constant, T is the absolute temperature and $[\ ]$ denotes the initial concentration of the substance. If, initially, the substances are at their molar concentrations, then expression (2.20) reduces to

$$\Delta F = RT \ln 1 - RT \ln K$$
$$= -RT \ln K,$$

and in this case we write

$$\Delta F^0 = -RT \ln K$$ (2.21)

where ΔF^0 is known as the *standard free energy change* for the reaction.

For some reactions the equilibrium constant cannot be directly measured, but it may be possible to calculate it by using the quantities known as *standard free energies of formation*. These are defined as follows:

From equation (2.19) for K we may write

$$RT \ln K = x RT \ln [X]_e + y RT \ln [Y]_e + + +$$
$$- a RT \ln [A]_e - b RT \ln [B]_e - - -$$ (2.22)

For each substance, the quantity defined as

$$\Delta F_f^0 = RT \ln [\ ]_e$$ (2.23)

is known as the standard free energy of formation.

This quantity is very useful since, if we do not know the reaction constant we can find it by looking up the tables of the standard free energies of formation of the participants.

Care must be exercised in using the values of ΔF_f° because a substance will have different values of ΔF_f° according to its particular state. Thus the value of ΔF_f° for a solid will be different for that of the substance in an unsaturated solution.

Example I

Suppose we have the reversible reaction

$$A + B \rightleftharpoons X + Y$$

at equilibrium and we suddenly increase 100-fold the concentration of A. Let us investigate the change in free energy.

The equilibrium constant is given by

$$K = \frac{[X]_e[Y]_e}{[A]_e[B]_e}.$$

The initial concentrations of all the participants except A are the equilibrium concentrations. The concentration of A is 100 times the equilibrium concentration. We may write the formula for ΔF as follows:

$$\Delta F = RT\ln \frac{[X]_e[Y]_e}{100[A]_e[B]_e}$$
$$- RT\ln \frac{[X]_e[Y]_e}{[A]_e[B]_e}$$

By making use of the rules for the manipulation of logarithms, we can cancel out nearly all the terms in the above expression so that it reduces to

$$\Delta F = RT\ln 1/100$$
$$= -RT\ln 100$$
$$= -2.303\, RT\log_{10} 100$$
$$= -4.606\, RT$$

From this example it is seen that, if the concentration of the reactants is increased, ΔF is negative and the reaction is exergonic, i.e. energy is given out as the reaction proceeds to equilibrium. A similar result is obtained if the concentration of the products is reduced.

On the other hand, if the concentration of the reactants is decreased or the concentration of the products is increased, then the reaction is endergonic.

Example II

In the reaction

$$\text{citrate}^{3-} \rightarrow \text{cis-aconitate}^{3-} + H_2O$$

the standard free energies of formation at $25°C$ are as follows:

citrate^{3-}	279.24 k.cals
cis-aconitate^{3-}	220.51 k.cals
H_2O	56.69 k.cals.

What is the reaction constant?

At equilibrium conditions we may write

$$K = \frac{[\text{cis-aconitate}^{3-}]_e\,[H_2O]_e}{[\text{citrat-e}^{3-}]_e}$$

$$RT\ln K = RT\ln [\text{cis-aconitate}^{3-}]_e$$
$$+ RT\ln [H_2O]_e$$
$$- RT\ln [\text{citrate}^{3-}]_e$$
$$= 220.51 + 56.69 - 279.24$$
$$= -2.04 \text{ k.cals} = -2040 \text{ cals}$$

Now $R = 1.987$ cal. $\deg^{-1}\text{mole}^{-1}$ and $T = 25°C = 298°$Abs.

so that
$$\ln K = - \frac{2040}{1.987 \times 298} = -3.4452$$

and

$$K = 0.0319.$$

3

GRAPHS AND CO-ORDINATE GEOMETRY

3.1 INTRODUCTION

The representation of experimental data by means of a graph is often a very useful means of showing at a glance how two quantities may be related, e.g. how the weight of a young animal increases from birth. Such a relationship may not be at all apparent if the various values are set out in tabular form. For instance, Table 3-I shows typical values of the weight of a human infant during its first year. The age is in months; the weight in pounds.

TABLE 3.I

Age	Weight	Age	Weight
birth	6.5	7	18.5
1	8.5	8	19.5
2	11.0	9	20.0
3	13.25	10	20.75
4	14.75	11	21.25
5	16.25	12	21.75
6	17.5		

In this table it is not easily seen if the infant increases its weight steadily with age or whether the increase is more rapid just after birth than it is say during the second six months of life. The values in the table are now plotted in the following graph (Fig. 3.1).

It is seen from this graph how the infant's weight increases from birth. Immediately it is noted that during the first three or four months the growth increases more rapidly than it does from about the eighth month onwards.

Usually graphs are plotted on squared paper and the quantity under the control of the experimenter, known as the independent variable, is plotted as *abscissa* i.e. in the horizontal direction, and the measured quantity, the dependent variable, is plotted as *ordinate,* i.e. in the vertical direction. In the case of the growth curve of the infant plotted above, time is the independent variable because we take the measurements of the infant's weight at times of our own choosing. If on the other hand, we recorded the age of the infant when its weight was say 10 lb, 12 lb,

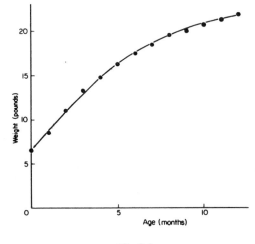

Fig. 3.1.

14 lb, etc. then weight would be the independent variable and time would be the dependent variable.

One exception to note is that when time is one of the quantities involved it is usually plotted as abscissa whether or not it is the independent variable.

It is very important when drawing graphs to indicate the scales that are used and to so choose them that the graph is able to show up points of interest to the best advantage.

The graph plotted above is produced from data which are obtained by taking measurements. It was not plotted according to a known mathematical relationship. In certain cases it is possible to plot a series of experimental results and then to deduce a mathematical formula to describe them. Such a formula is said to be *empirical*—it is derived from experimental observations. The process of curve fitting i.e. deducing the appropriate formula can sometimes be very difficult. Usually, the ultimate goal in curve fitting is to manipulate the results so that they can be plotted in the form of a straight line.

On the other hand, it is possible that the results do fit a curve which has been derived according to a previously set-up theory. For instance, the growth curve for bacteria can be calculated and experiments performed to see if the observed values do indeed fit the calculated curve.

3.2   CARTESIAN CO-ORDINATES

Consider a set of rectangular axes OX and OY intersecting at O. If OM = x and MP = y, then P is called the point (x,y), x is called the abscissa of P and y is called the ordinate of P.

Thus x and y are the co-ordinates of the point P and they fix its position in the plane XOY.

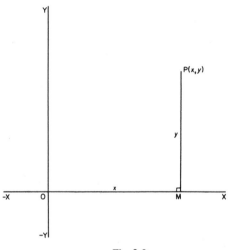

Fig. 3.2.

In Fig. 3.2 the co-ordinates of P have positive values. This is not always the case because the axes divide the plane XOY into four quadrants. This is better shown in Fig. 3.3.

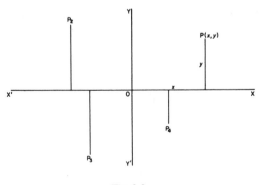

Fig. 3.3.

In this figure XO is produced backwards to form the line OX′ and YO is produced backwards to form the line OY′. Distances measured along OX from O and distances along OY from O are reckoned positive. Distances measured along OX′ from O and along OY′ from O are reckoned negative.

Thus the point P_2 in the figure, which lies in the quadrant X′OY has a negative abscissa and a positive ordinate.

The point P_3 which lies in the quadrant X′OY′ has a negative abscissa and a negative ordinate.

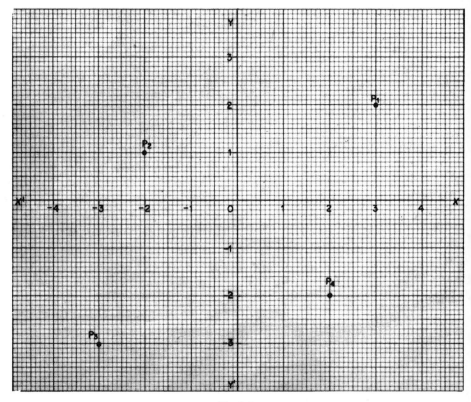

Fig. 3.4.

The point P_4 which lies in the quadrant Y′OX has a positive abscissa and a negative ordinate.

Now let us consider as examples four points P_1, P_2, P_3, P_4 whose co-ordinates are given by $(3, 2)$, $(-2, 1)$, $(-3, -3)$, and $(2, -2)$. These are plotted on the graph paper (Fig. 3.4). In order to do this we first of all draw in the co-ordinates with the origin O roughly at the centre of the paper. We now number off the squares along the axes using positive numbers along OX and OY and negative numbers along OX′ and OY′.

To plot the point P_1 (3, 2) measure a distance 3 along OX and then a distance 2 parallel to OY. We thus arrive at the point P_1 and it is plotted as shown.

To plot the point P_2 (−2, 1) measure a distance 2 along OX′ and then a distance 1 parallel to OY.

To plot the point P_3 (−3, −3) measure a distance 3 along OX′ and then a distance 3 parallel to OY′.

To plot the point P_4 (2, −2) measure a distance 2 along OX and then a distance 2 parallel to OY′.

If one of the co-ordinates of a point is zero, the point lies on one of the axes. If the ordinate is zero the point will lie on the X-axis, i.e. on the line X′OX. If the abscissa is zero the point will lie on the Y-axis, i.e. on the line Y′YO.

The point with both co-ordinates equal to zero, i.e. (0, 0) lies on both the X-axis and the Y-axis and this point is, of course, the origin O.

Exercise 3.2

The following sets of points have the co-ordinates shown. Plot each set of points on a separate sheet of graph paper and choose appropriate scales.

(1) (7, −2), (−3, 3), (3, −2), (4, 0)
(2) (30, 0), (−90, 30), (45, 45), (10, −70)
(3) (17, 1), (−10, −3), (22, 7), (5, −5)
(4) (0.01, 0.15), (0.02, −0.08), (0.20, 0.15).

3.3 THE LENGTH OF A LINE JOINING TWO POINTS

The distance between two points in a plane may be derived as follows. Consider the two points P_1 (x_1, y_1) and P_2 (x_2, y_2) [Fig. 3.5].

By drawing P_1M and P_2N parallel to the Y-axis and P_1C parallel to the X-axis it can be seen from the resultant figure that since

$$P_2 N = y_2,$$

and

$$P_1 M = y_1$$

then

$$P_2 C = (y_2 - y_1).$$

Also since,

$$ON = x_2,$$

$$OM = x_1$$

then

$$P_1 C = MN = x_2 - x_1.$$

By Pythagoras' theorem

$$P_1 P_2^2 = P_2 C^2 + P_1 C^2$$

and thus $$P_1 P_2 = [(y_2 - y_1)^2 + (x_2 - x_1)^2]^{1/2} \qquad (3.1)$$

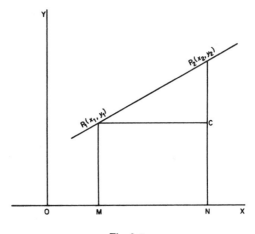

Fig. 3.5.

For example, if we have two points whose co-ordinates are $(3,2)$ and $(1,-7)$, then the distance between them is given by

$$[(-7 - 2)^2 + (1 - 3)^2]^{1/2}$$
$$= [(-9)^2 + (-2)^2]^{1/2}$$
$$= [81 + 4]^{1/2}$$
$$= [85]^{1/2} \simeq 9.22$$

From equation (3.1) we see that the distance from the origin of a point $P_1 (x_1, y_1)$ is given by $[y_1^2 + x_1^2]^{1/2}$.

This result is obtained by putting x_2 and y_2 equal to zero in expression (3.1.)

Expression (3.1) for the distance between two points is used when it is required to determine the distance between two points on a microscope slide. The set of axes is provided by an attachment which is fitted to the stage of a microscope. The slide is moved in two directions at right angles to one another by means of two screw threads which have vernier scales. The point of interest on the slide is centred up beneath a pair of cross wires in the microscope and the X and Y vernier readings are noted. A second point of interest is next centred up and its vernier readings are noted. The distance between the points is then calculated according to expression (3.1). This provides a particularly useful method for measuring the thickness of blood vessels or the size of a small organ.

Exercises 3.3

(1) Calculate the distances between the following pairs of points.

  (a) $(7.2, 3.1)\,(2.8, 1.9)$
  (b) $(6, -4)\,(-3, -2)$
  (c) $(0.1, 0.1)\,(1.1, 0.7)$
  (d) $(21, 13)\,(18, 17)$

(2) A microscope slide shows an artery which is nearly circular in cross section. In order to measure the diameter of the artery vernier readings were taken at opposite ends of a diameter. The readings for the two points expressed in mm were $(12.8, 62.3)$ and $(10.0, 65.0)$. Calculate the diameter of the artery.

3.4 THE SLOPE OF A LINE JOINING TWO POINTS

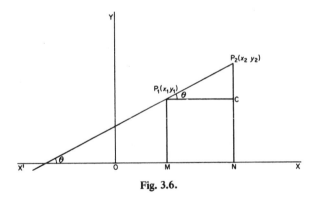

Fig. 3.6.

The line joining the two points $P_1\,(x_1, y_1)$ and $P_2\,(x_2, y_2)$ will intersect with the X-axis at an angle θ. The slope of the line is defined as $\tan\theta$. From Fig. 3.6 it is seen that

$$\tan\theta = \tan P_2\hat{P_1}C = \frac{(y_2 - y_1)}{(x_2 - x_1)} \tag{3.2}$$

For example, the slope of the line joining the two points whose co-ordinates are $(2, 7)$ and $(-1, 5)$ is

$$\frac{(5-7)}{(-1-2)} = \frac{-2}{-3} = \frac{2}{3}$$

If θ is greater than $90°$, but less than $180°$, $\tan\theta$ has a negative value. For example, the slope of the line joining the two points in Fig. 3.7 is negative.

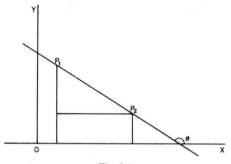

Fig. 3.7.

In this case, the application of the formula (3.2) to find $\tan\theta$ will yield a negative value. For instance, the slope of the line joining the points $(2, 7)$ and $(7, 2)$ is given by

$$\tan\theta = \frac{(2-7)}{(7-2)} = \frac{-5}{5} = -1.$$

At this point it is perhaps a good idea to remind ourselves of the value of the tangents of certain angles. Consider the triangle ABC Fig. 3.8 which is right-angled at C.

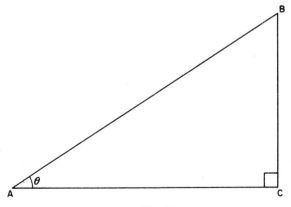

Fig. 3.8.

By definition tan $B\hat{A}C$ = BC/AC. Consider now the following cases:

(a) BAC = $0°$ i.e. BC = 0

$\quad$ tan 0 = O/AC = 0. $\hspace{6cm}$ (3.3)

(b) $B\hat{A}C = 30°$

In this case the lengths of the sides are in the ratio $2:\sqrt{3}:1$ so that

$\quad$ tan 30 = $1/\sqrt{3}$ $\hspace{6cm}$ (3.4)

(c) $B\hat{A}C = 45°$

In this case BC = AC so that

$\quad$ tan 45 = 1 $\hspace{6.5cm}$ (3.5)

(d) $B\hat{A}C = 60°$

In this case, as in case (b), the sides of the triangle are in the ratio $2:\sqrt{3}:1$ but BC is larger than AC so that

$\quad$ tan 60 = $\sqrt{3}/1 = \sqrt{3}$ $\hspace{5cm}$ (3.6)

(e) $B\hat{A}C = 90°$

In this case the sides BC and AB are of infinite length and therefore

$\quad$ tan 90 = ∞/AC = ∞ $\hspace{5.5cm}$ (3.7)

(f) Angles greater than $90°$

In this case we cannot refer to the triangle ABC above, but the tangents of angles greater than $90°$ are considered as follows: Suppose a radius OP, starting from an initial position OX is rotated in an anticlockwise direction. Some possible positions of OP after it has swept out the angle θ are shown in Fig. 3.9.

$\quad$ In (i) the angle θ is less than $90°$ and tan θ is given, according to the above definition by

$\quad$ tan $\theta = y/x$

In the other examples the definition for tan θ holds provided appropriate signs are attached to the abscissa and ordinate of the point P according to its position in the diagram.

Thus in case (ii)

$\quad$ tan $\theta = y/(-x) = -y/x$.

In case (iii)

$\quad$ tan $\theta = (-y)/(-x) = y/x$.

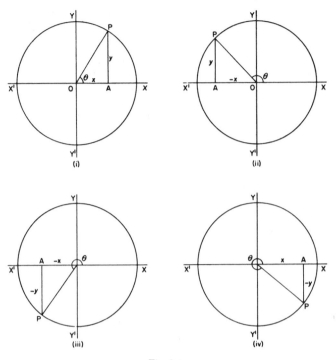

Fig. 3.9.

In case (iv)

$$\tan \theta = -y/x.$$

In other words tan θ is positive if 0 lies in the first and third quadrants and negative if it lies in the second and fourth quadrants.

Sometimes it is useful to know the slope of a line which is perpendicular to a line of known slope m.

In Fig. 3.10

$$\hat{POB} = \theta$$
$$\tan \theta = m,$$
$$\hat{BOQ} = 90° + \theta$$

Since $\hat{QAO} = 90°$,
then $\hat{AQO} = \theta$
Also $\tan \hat{BOQ} = -QA/AO$,
but $\tan \theta = AO/QA = m$
Therefore

$$\tan \hat{BOQ} = \tan (90° + \theta) = -1/m$$

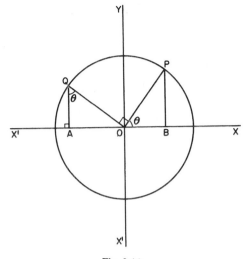

Fig. 3.10.

The tables at the end of this book give values of the tangents of angles up to 90°. They can be used to determine the tangents of all angles between 0° and 90° at intervals of 1'. For angles of an exact number of degrees the value of the tangent is given in the first column of figures. The next nine columns give the tangents of angles which progressively increase by 6'. For example, the tables give

$$\tan 19° = 0.3443$$
$$\tan 19°6' = 0.3463$$
$$\tan 19°12' = 0.3482$$

The five columns on the right of the page are headed mean differences. These values are the increments to be added to the last places of decimals to give tangents of angles which lie within the 6' intervals.

For instance.

$$\tan 19°6' = \underline{0.3463}$$
$$\tan 19°7' = 0.3463 + 0.0003 = \underline{0.3466}$$
$$\tan 19°8' = 0.3463 + 0.0007 = \underline{0.3470}.$$

Notice that in the tables the values of the tangents of angles between 84° and 90° increase at such a rate that the mean differences cease to be sufficiently accurate.

Tangents of angles greater than 90° may be found by using the tables, provided that we remember to give the correct sign to the values.

(1) Use the tables to find the tangents of the following angles:

    (a) 19°37'   (b) 192°29'
    (c) 33°10'   (d) 282°33'
    (e) 171°31'   (f) 350°
    (g) 213°10'   (h) 100°30'

(2) Use the tables to find the angles whose tangents have the following values:

    (a) 17.89   (b) 0.5895
    (c) −1.4290   (d) 0.8098
    (e) −1.7723   (f) 0.3584

(3) Find the tangents of the angles to the X-axis of the straight lines which pass through the following pairs of points:

    (a) $(7.2, -3.1), (4.9, 6.2)$
    (b) $(0.3, 0.3), (1.9, 2.7)$
    (c) $(13, 17), (-17, -13)$
    (d) $(109, 67), (25, 0)$
    (e) $(-2, 1), (-3, 1)$.

3.5   THE STRAIGHT LINE

The simplest curve is the straight line and this is represented by an equation in which all the terms in x and y are of degree 0 or 1. For example, the following equations represent straight lines:

$$y = mx + C \qquad\qquad (3.8)$$

$$x + my + n = 0 \qquad\qquad (3.9)$$

$$\frac{x}{a} + \frac{y}{b} = 1 \qquad\qquad (3.10)$$

$$x = a \qquad\qquad (3.11)$$

$$y = b. \qquad\qquad (3.12)$$

In these equations all terms except x and y are constants. Such equations are known as first degree equations. If any equation contains a term with a fractional degree of x or y, or a negative degree, or a degree greater than one, then the equation is non-linear.

Now let us consider several ways of determining the equation of a straight line.

(a) The equation $y = mx + C$.

Suppose that it is given that a straight line has a slope m and makes an intercept C on the Y-axis. This information is sufficient to uniquely determine the straight line or, in other words, there is only one straight line which can have a slope m and intercept C. This straight line is represented in Fig. 3.11

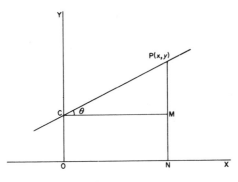

Fig. 3.11.

The line makes an angle θ with the X-axis such that $\tan \theta = m$. Let a point P with co-ordinates (x, y) be any point on this line. From P drop a perpendicular to meet the X-axis in N and from C drop a perpendicular to meet PN in M.
Now

$$m = \tan \theta = PM/CM = (y - C)/x.$$

Rearrangement of this gives

$$y - C = mx$$

or

$$y = mx + C.$$

The equation $y = mx + C$ therefore represents a straight line of slope m making an intercept C on the Y-axis. Thus if, for example, we see the equation

$$y = 3x + 6$$

we can say immediately that this equation represents a straight line of slope 3 which makes an intercept 6 on the Y-axis.

To obtain the slope of any straight line always rearrange the equation in the form $y = mx + C$. For instance the equation

$$6x - 3y + 10 = 0$$

represents a straight line since it is in the form of equation (3.9) above. We can, however, divide throughout by 3 and rearrange the equation to give

$$y = 2x + 3\tfrac{1}{3}$$

which is a line of slope 2 making an intercept $3\tfrac{1}{3}$ with the Y-axis.

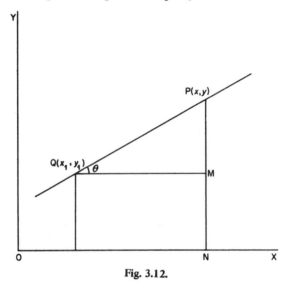

Fig. 3.12.

If, in the equation $y = mx + C$ the intercept C is zero then the straight line $y = mx$ passes through the origin: if m the slope is zero, then the equation $y = C$ represents a straight line parallel to the X-axis and distant C from it.

The straight line $y = mx + C$ crosses the X-axis at a point obtained by putting $y = 0$ in the equation.

Thus $0 = mx + C$ gives $x = -C/m$

The line $y = mx + C$ makes an intercept $-C/m$ on the X-axis.

(b) The equation $\qquad\qquad y - y_1 = m(x - x_1)$ $\qquad\qquad$ (3.13)

Suppose that we are given that a straight line has a slope m and passes through a given point Q (x_1, y_1). Let $P(x, y)$ be any other point on this line (Fig. 3.12). From P drop a perpendicular to meet the X-axis in N and from Q drop a perpendicular to meet PN in M. Now

$$\tan \theta = m = \text{PM/QM} = (y - y_1)/(x - x_1).$$

Rearrangement of this equation gives

$$y - y_1 = m(x - x_1).$$

This equation represents a straight line of slope m which passes through the point (x_1, y_1)

Example I

Find the equation of the line which has a slope $-\frac{1}{3}$ and passes through the point $(-3, 2)$.
The given values are fitted in the equation

$$y - y_1 = m(x - x_1)$$

to give

$$y - 2 = -\frac{1}{3}(x + 3)$$

which is rearranged to give

$$x + 3y - 3 = 0$$

or

$$y = -\frac{1}{3}x + 1.$$

The resultant equation is seen to make an intercept 1 with the Y-axis.

(c) The equation $\qquad \dfrac{y - y_1}{x - x_1} = \dfrac{y_2 - y_1}{x_2 - x_1} \qquad$ (3.14)

Two points are sufficient to define a straight line. Let Q (x_1, y_1) and R (x_2, y_2) be two given points on a straight line. Let P (x, y) be any other point of the line. By dropping perpendiculars as shown in Fig. 3.13 we have

$$\tan \theta = RS/QS = PT/QT$$

so that

$$\frac{y_2 - y_1}{x_2 - x_1} = \frac{y - y_1}{x - x_1}$$

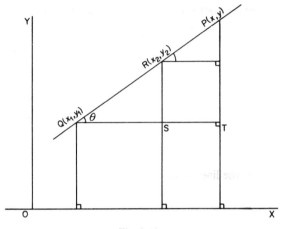

Fig. 3.13.

Example II

To determine the equation of the straight line which passes through $(3,4)$ and $(6,2)$. The co-ordinates of the points are inserted into equation (3.14) to give

$$\frac{y-4}{x-3} = \frac{2-4}{6-3} = -\frac{2}{3}$$

or

$$y = -\tfrac{2}{3}\,x + 6.$$

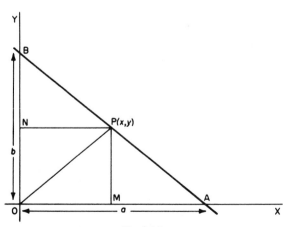

Fig. 3.14.

The required line makes an intercept 6 on the Y-axis and has a slope $-\frac{2}{3}$

(d) The equation $\qquad\qquad \dfrac{x}{a} + \dfrac{y}{b} = 1$ $\qquad\qquad\qquad$ (3.15)

This equation is known as the intercept equation. Consider a straight line making intercepts a and b on the X and Y-axes respectively. Let $P(x,y)$, [Fig. 3.14] be any point on the line. From P drop a perpendicular to meet the X-axis at M and the Y-axis at N. Join PO. Considering the triangles BOA, POB and POA we may write

$$\text{Area POA} + \text{Area POB} = \text{Area BOA}$$

$$\frac{a \times y}{2} + \frac{b \times x}{2} = \frac{a \times b}{2}$$

This equation may be rearranged to the form

$$\frac{x}{a} + \frac{y}{b} = 1.$$

Example III

A straight line makes an intercept 6 on the X-axis and 4 on the Y-axis. Its equation is given by

$$\frac{x}{6} + \frac{y}{4} = 1$$

or

$$2x + 3y = 12$$

Exercise 3.5

(1) Find the equations of the lines passing through the following pairs of points:

(a) $(3,2); (1,0)$ $\qquad$ (b) $(a, b); (-b, a)$
(c) $(4,2); (-8,-2)$ $\qquad$ (d) $(3, 1); (2, 2)$
(e) $(0,3); (2,0)$ $\qquad$ (f) $(3,2); (-2,-3)$

(2) Find the equation of the lines

    (a) passing through $(3,9)$ with slope 2
    (b) passing through $(1,2)$ with slope 1
    (c) passing through $(-1,0)$ with slope 3
    (d) passing through $(-2,3)$ with slope -1.

(3) Derive the equation of the straight line of slope m which makes an intercept d on the X-axis.

3.6   SPECIAL FORMS OF THE STRAIGHT LINE

(a) Lines parallel to the axes.

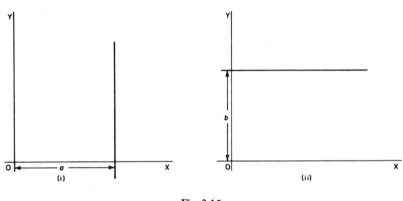

Fig. 3.15.

Consider the straight lines in Fig. 3.15. In case (i) the line cuts the X-axis at a distance a from the origin. Since the line is drawn parallel to the Y-axis, the intercept on the Y-axis is at infinity. We want to find the equation of this line. The intercept equation

$$\frac{x}{a} + \frac{y}{b} = 1$$

may be applied but, since $b = \infty$ and therefore $y/\infty = 0$, the equation reduces to,

$$\frac{x}{a} = 1 \text{ or } x = a$$

Likewise, we may show that a straight line parallel to the X-axis having an intercept b on the Y-axis Fig. 3.15 (ii) is given by

$$\frac{y}{b} = 1 \text{ or } y = b$$

If the values of a or b in the above equations are zero, we have equations of the Y-axis and X-axis respectively.

(b) Lines through the origin,

If, in the equation $y = mx + C$, the value of C is zero, then the equation

$$y = mx$$

represents a straight line of slope m which passes through the origin.

3.7 THE FITTING OF DATA TO A STRAIGHT LINE

Suppose that we have a set of values of the quantity y for corresponding values of the quantity x and we suspect that there is a linear relationship between x and y of the form

$$y = mx + C$$

As an example we consider the following series of values

x	0	1	2	3	4	5	6	7	8
y	2.0	2.4	2.75	3.1	3.5	3.9	4.25	4.6	5.0

If these points are plotted on graph paper as in Fig. 3.16 it is seen that they can be joined fairly easily by a straight line. The slope and intercept may be measured and it is found that the slope is $\frac{3}{8}$ and the intercept is 2. Therefore the equation of this straight line is

$$y = \frac{3}{8} x + 2$$

or

$$8y = 3x + 16.$$

Now the example given above is such that the drawing of the straight line is quite easy and that all the points lie on or very nearly on the straight line. Sometimes, however, especially where experimental data are concerned, one can get the very

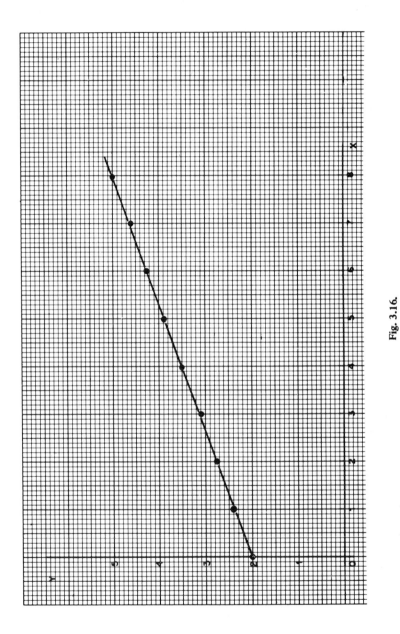

Fig. 3.16.

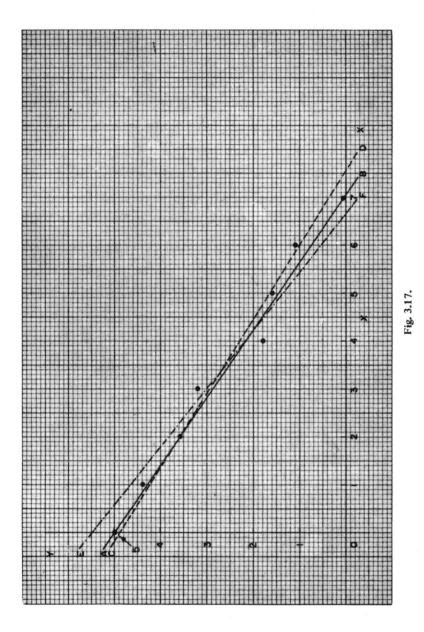

Fig. 3.17.

strong impression that a linear relationship exists but that it is very difficult to determine exactly by eye the position of the straight line. Consider for example the following set of points:

x	0	1	2	3	4	5	6	7
y	5	4.4	3.6	3.2	1.8	1.6	1.1	0.1

These are plotted on the graph Fig. 3.17. A cursory inspection shows that these points lie on or near a straight line but the exact position of the line will be a matter of personal judgment. One person might draw in the line AB so that it passes through the first point and the last point. The line will make an intercept of 5 on the Y-axis and 7.1 on the X-axis. Now according to the intercept equation this will give a line

$$\frac{x}{7.1} + \frac{y}{5} = 1$$

or

$$y = -\frac{5}{7.1}x + 5.$$

Another person might decide that the line AB is not the most accurate line because more points lie above this line than below it. He might decide that the line CD gives a better fit. This line has intercepts of 4.9 and 7.6 to give an equation

$$\frac{x}{7.6} + \frac{y}{4.9} = 1$$

or

$$y = -\frac{4.9}{7.6}x + 4.9.$$

Yet a third person might decide that the middle points might be more important than the end points and draw in the line EF to give the equation

$$\frac{x}{6.7} + \frac{y}{5.5} = 1$$

or

$$y = -\frac{5.5}{6.7} x + 5.5.$$

It is thus seen that three people might draw in lines to these points which have slopes of $-5/7.1 = -0.70$, $-4.9/7.6 = -0.64$ and $-5.5/6.7 = -0.82$. Clearly then we must look for some sort of method which we can use to ensure that the line that we draw in to fit the points is the most accurate for the data available. Methods for doing this are given in the next sections.

Exercises 3.7

The following sets of points are good fits to straight lines. Plot the points, draw in the straight lines and measure the slopes and intercepts on the Y-axis.

(a)

x	1	3	5	7	9
y	1.3	2.1	2.8	3.6	4.4

(b)

x	1	2	3	4	5	6
y	4.5	4.0	3.4	3.0	2.6	2.0

3.8   METHOD OF AVERAGES

Consider any pair of values of x_n the variable and y_n the corresponding experimental value. If the relationship

$$y = mx + C$$

holds and if the point (x_n, y_n) lies on this line we have the relationship

$$y_n - mx_n - C = 0.$$

If on the other hand the point lies off the line we have

$$y_n - mx_n - C = r_n \tag{3.16}$$

where r_n is termed a 'residual' and represents the vertical distance of the plotted point above the straight line. The residual has a positive or a negative value according to whether the plotted point lies above or below the straight line. If all the plotted points are considered the straight line to which they approximate is considered to be such that the sum of the residuals is zero.

Thus for n plotted points we write

$$y_1 - mx_1 - C + y_2 - mx_2 - C + \ldots$$
$$+ y_m - mx_n - C = 0$$

We may rewrite this equation in the form:

$$(y_1 + y_2 + \ldots y_n) - m(x_1 + x_2 \ldots + x_n)$$
$$- nC = 0 \qquad (3.17)$$

or

(the sum of the values of y) −
(the sum of the values of x) $- nC = 0$

We are of course unable to determine the values of m and C from this equation because it contains *two* unknown quantities. What we do in fact is to divide the plotted points into roughly two groups and make two separate equations of the form (3.17). The two equations would enable us to calculate the values of m and C. In order to clarify the procedure let us return to the set of points quoted in Section 3.7. This is divided into two groups and the sum of the values of x and the sum of the values of y are found for each group.

TABLE 3.II

Group I		Group II	
x	y	x	y
0	5.0	4	1.8
1	4.4	5	1.6
2	3.6	6	1.1
3	3.2	7	0.1
6	16.2	22	4.6

Since there are four values in each group we may write out the two equations.

$$16.2 - 6m - 4C = 0$$
$$4.6 - 22m - 4C = 0$$

The second equation is subtracted from the first to give

$$16m = -11.6$$
$$m = -0.725.$$

Substitution of this value in either equation yields a value

$$C = 5.14$$

The equation to the required line is thus

$$y = -0.725x + 5.14$$

3.9 METHOD OF LEAST SQUARES

Another method of determining the straight line through a set of points is known as the method of least squares. This method is somewhat more complicated to use than the above method but it is easily carried out if a desk calculating machine is available. Before describing it a type of mathematical shorthand is introduced in order to make the description neater. Consider again equation (3.17). Instead of writing 'the sum of the values of y' we write Σy where Σ—the Greek letter capital sigma—is the shorthand way of saying 'the sum of the values of'. Likewise we write Σx to mean 'the sum of the values of x' so that equation (3.17) may be written

$$\Sigma y - m\Sigma x - nC = 0 \tag{3.18}$$

The method of least squares also involves an equation of this form but in this case the Σy and Σx refer to the sums of *all* the values of y and x respectively. This equation contains the two unknowns m and C. The second equation necessary to determine their values is obtained as follows:

The equation of the straight line is of the form

$$y = mx + C.$$

Now multiply throughout by x to obtain

$$xy = mx^2 + Cx$$

Now if a plotted point (x_n, y_n) lies near to the straight line we have

$$x_n y_n - mx_n - Cx_n = P_n$$

where P_n is similar to the residual mentioned in the previous section. The line of best fit is reckoned to be such that, when all the plotted points are considered, the sum of the values of the residuals is zero. Thus for n points:

$$(x_1 y_1 + x_2 y_2 + + + x_n y_n) - m(x_1^2 + x_2^2 + + + + + + + x_n^2)$$
$$- C(x_1 + x_2 + + + + x_n) = 0$$

Using the notation explained above we write

$$\sum xy - m \sum x^2 - C \sum x = 0$$

$\sum xy$, $\sum x^2$ and $\sum x$ can be calculated from the given data so that we can get a second equation containing m and C.

As an example let us again use the set of data that was used in the previous section. First of all we draw up the following table of values:

TABLE 3.III

x	y	xy	x^2
0	5	0	0
1	4.4	4.4	1
2	3.6	7.2	4
3	3.2	9.6	9
4	1.8	7.2	16
5	1.6	8.0	25
6	1.1	6.6	36
7	0.1	0.7	49

$$\sum x = 28 \quad \sum y = 20.8 \quad \sum xy = 43.7 \quad \sum x^2 = 140$$

The number of plotted points is 8, so that inserting the values of the quantities in (3.18) gives

$$20.8 - 28m - 8C = 0$$

and from (3.19) we likewise obtain

$$43.7 - 140m - 28C = 0$$

If we divide the first equation by 2 and then multiply by 7 we get

$$72.8 - 98m - 28C = 0$$

We now subtract one equation from the other to get

$$29.1 + 42\,m = 0$$

$$m = -\frac{29.1}{42} = -0.693 \text{ (to three figures)}$$

and hence

$$C = 5.02 \text{ (to three figures)}.$$

The values of m and C as obtained by the method of least squares differ slightly from the values obtained by the method of averages. There are more than one ways of deciding the position of the 'best line' and some methods are more accurate than others. For reasons which cannot be explained within the scope of this book, it is reckoned that the method of least squares is more accurate than the method of averages.

Exercises 3.9

The following experimental values of y for various values of x were obtained in an experiment. Plot these points and find the equation of the best straight line as determined by both the method of averages and the method of least squares.

x	0	1	2	3	4	5	6	7
y	4.6	4.1	3.4	2.9	2.5	2.0	1.4	0.9

3.10 REGRESSION LINES

In the last section we considered how to draw a straight line through a set of points by means of the method of least squares. Two equations were used in the determination of m and C, the constants of the required line. They were

$$\sum y = m \sum x + nC$$
$$\sum xy = m \sum x^2 + C \sum x$$

We now divide the equations by n and get

$$\frac{\sum y}{n} = \frac{m \sum x}{n} + C$$

$$\frac{\sum xy}{n} = \frac{m \sum x^2}{n} + \frac{C \sum x}{n}$$

These equations can be written in a shorthand form

$$\bar{y} = m\bar{x} + C.$$

$$\frac{\sum xy}{n} = \frac{m \sum x^2}{n} + C\bar{x}$$

where $\bar{y} = \sum y/n$, and $\bar{x} = \sum x/n$ are the mean values of y and x respectively. If we multiply the first equation by $\bar{x}$ and then subtract the second equation we obtain

$$\bar{x}\bar{y} - \frac{\sum xy}{n} = m\bar{x}^2 - \frac{m \sum x^2}{n}$$

which gives

$$m = \frac{\sum xy - n\bar{x}\bar{y}}{\sum x^2 - n\bar{x}^2}.$$

If we now substitute this value of m into either of the original equations we obtain

$$C = \frac{\bar{y} \sum x^2 - \bar{x} \sum xy}{\sum x^2 - n\bar{x}^2}.$$

The line that we have obtained by this method is known as a *regression line*. This line in particular gives an estimate of values of y from values of x and the line is called the regression line of y on x.

Sometimes of course we may need to estimate values of x from values of y. In this case we calculate the regression line of x on y. Such a line has the equation

$$x = m'y + C'$$

where

$$m' = \frac{\sum xy - n\bar{x}\bar{y}}{\sum y^2 - n\bar{y}^2} \qquad C' = \frac{\bar{x} \sum y^2 - \bar{y} \sum xy}{\sum y^2 - n\bar{y}^2}$$

Only if all the points lie exactly on a straight line will the two lines of regression be identical. The reason for this is that, in calculating the line of regression of y on x, the vertical deviations of these points from the line are minimised. In the calculation of the line of regression of x on y the horizontal deviations of the points from the line are minimised.

Exercises 3.10

(1) Derive the formulae for m' and C' quoted above for the line of regression of x on y.
(2) Use the values given in Exercise 3.9 to calculate the line of regression of x on y.

3.11   CORRELATION COEFFICIENTS

From the foregoing sections it is seen that, although the formulae derived are intended to produce a straight line through a set of points, any set of points may be used in these formulae. A problem arises then as to *how well* the set of points lie on the straight line. We need some measure of the degree of linearity between x and y. This is achieved by calculating the *coefficient of correlation* between two variables x and y and it is defined as

$$r = \frac{\sum xy - n\bar{x}\bar{y}}{[\sum x^2 - n\bar{x}^2]^{1/2} [\sum y^2 - n\bar{y}^2]^{1/2}}$$

By reference to the formulae of the previous section we can show that

$$r = [m.m']^{1/2}.$$

Now whatever the values of the x's and y's it can be shown, by reference to more advanced textbooks on statistics, that r will always have values between $+1$ and -1. If r is positive it indicates that the higher values of x will have correspondingly higher values of y. If r is negative the lower values of x will have higher values of y and there is then said to be an *inverse* relationship between x and y.

The calculation of the coefficient of correlation is quite straightforward provided that there are only a few values of x and y involved. Otherwise a desk calculator or even a computer must be used.

However, let us assume that we have the facilities and are able to calculate the coefficient of correlation. We next consider the meaning of the value we obtain as an indication of the degree of linearity. It will be very unlikely that our set of points will lie exactly on a straight line. If this situation did exist then r would

take the value +1 for a direct relationship and −1 for an indirect relationship.
All other situations will yield values of r which lie somewhere between +1 and
−1, for an indirect relationship. All other situations will yield values of r which
lie somewhere between +1 and −1. If then, the points do not lie exactly on a
straight line, can we say that in fact there is any linear relationship? What we
have to do is to apply some sort of test and ask the question 'what is the probability that a linear relationship exists?'. It is usually accepted that a linear relationship exists if the probability that the same result could be obtained by a pure
chance distribution of the points is less than 5%, or in other words, if we apply

TABLE 3.IV

n	r	n	r
6	0.82	40	0.31
8	0.71	50	0.28
10	0.63	60	0.254
15	0.52	70	0.235
20	0.44	80	0.22
30	0.36	100	0.20

a test of significance and get a 95% or more probability that there is a linear
relationship, then we accept that a linear relationship exists. If there is no linearity between a set of points then it is possible that a small number of points
might tend to show a correlation, but a large number of points would not show
a correlation. The reason for this is that for a few points a pure chance distribution might follow a straight line but this would not be the case for a large
number of points. Hence for a small number of points the value of r must be
much nearer to +1 or −1 than for a large number of points.

Table 3.IV gives values of r (or −r) for various values of n which must be
exceeded in order that a 95% probability of a linear relationship is established.
It is beyond the scope of this book to give the derivation of these values.

Exercises 3.11

Using the data given for the example of Exercise 3.9 determine the coefficient
of correlation. What can you deduce about the relationship of the values of x to
the values of y?

3.12 INFORMATION OBTAINABLE FROM GRAPHS

As mentioned at the beginning of this chapter, the purpose of drawing a graph is to show whether a series of experimental results follow a mathematical relationship and if so, such a relationship will be more apparent if the results are plotted on a graph than if the results are merely displayed in tabular form. In the last two sections we discussed how to calculate the best straight line to fit a set of results and in particular to calculate the value of the constants m and C. In the discussion of our results we must be very careful to express them in the correct units. So far we have talked of the variables x and y only in terms of numbers. But of course in much scientific work the values that are plotted will have definite units assigned to them. In addition to the familiar units of mass, e.g. gm, lb, tons; length, e.g. ft, cm; time, e.g. sec, h, we might also want to express our quantities in the units of volume, force, concentration weight etc. Often we might use units which have no common name.

All the units that we use will either be *dimensionless* or have dimensions which can be expressed as a particular combination of mass, length and time. Quantities which have dimensionless units include pure numbers, e.g. population size, or temperature. When we consider the dimensions of a quantity we write L for length, M for mass, and T for time. From these three fundamental units we can get a great variety of derived units. For instance area, which is obtained by multiplying a length by a length, has the dimension L^2; volume has the dimension L^3. We will now consider the dimensions of some of the familiar quantities that we are likely to encounter:

Density. Density is usually expressed as the number of pounds per cubic foot, or the number of grammes per cubic centimetre. We can say in fact that density is mass per unit volume and will therefore have the dimensions

$$\frac{M}{L^3} \text{ or } M.L.^{-3}$$

Velocity or speed. Miles per hour or centimetres per second etc. Dimensions

$$\frac{L}{T} \text{ or } L.T.^{-1}$$

Acceleration. Feet per second per second or centimetres per second per second etc. Dimensions $L.T.^{-2}$.

When dealing with the straight line equation the dimensions of the intercept C will be the same as the dimensions of the ordinate. The dimensions of the slope m will be the dimensions of the ordinate divided by the dimensions of the abscissa. For example, if we draw a graph of the passage of a single action potential along a nerve fibre we plot distance travelled as ordinate and time taken to reach the distance as abscissa. The plot of the points produces a straight line graph, the slope of which will have the dimensions of length/time, i.e. L/T. These are the dimensions of velocity. The slope of the line will give the value for the velocity of the impulse along the nerve fibre.

4

NON-LINEAR RELATIONSHIPS

4.1 THE POWER LAW

In this chapter we shall be considering certain non-linear relationships which are often of use to biologists. In this section we shall consider

$$y = Ax^n \tag{4.1}$$

where A and n are constants. Such a relationship is known as a *power law.*

If we have a set of data which we think obey the power law we would want to know the values of the constants A and n. If we just plot the data on ordinary graph paper we shall encounter the difficulties of not being able to decide which is the best curve to fit the points and the determination of the values of the constants from the graph will be rather haphazard. Techniques are available to enable us to overcome these difficulties. Consider equation (4.1). From our knowledge of logarithms we may write it in the form

$$\log y = \log A + n \log x \tag{4.2}$$

This can further be arranged in the form

$$Y = n\,X + \log A \tag{4.3}$$

where $X = \log x$ and $Y = \log y$. Also, since A is a constant, then $\log A$ is also a constant. Equation (4.3) is thus the equation of a straight line; instead of plotting the values of our data in the usual manner, we plot $\log y$ against corresponding values of $\log x$. If the power law is obeyed, the points will lie on a straight line. The line of best fit can be determined by methods described in the previous chapter. Having determined the line of best fit we can now determine the constants, n is the slope of the straight line and the intercept gives $\log A$.

Before we go on to discuss the power law by way of examples, let us consider
how the shapes of the curves of the power law equation differ according to the
values of the constant n. In each case to be considered the constant A is taken as
unity.

(a) n is positive and greater than unity.
For $x = 0, x^n = 0$.
For x positive but less than 1, x^n is less than x.
For $x = 1, x^n = 1$.
For x greater than 1, x^n exceeds x.

A typical curve, in this case for $n = 2$, is shown in Fig. 4.1a.

(b) $n = 1$.

This is the trivial case $y = x$ which is a straight line through the origin.

(c) n is positive but less than 1.
For $x = 0, x^n = 0$.
For x positive but less than unity, x^n exceeds x.
For x positive but greater than unity, x^n is less than x.

A typical curve, in this case for $n = \frac{1}{2}$, is given in Fig. 4.1b.

(d) n is negative.

In the case of a negative power, the power function may be written

$$y = x^{-m} \text{ or } y = \frac{1}{x^m}$$

where m is positive.

For $x = 0, y = \infty$.
For x greater than 0 but less than 1, x^{-m} exceeds 1.
For x greater than 1, x^{-m} is less than 1.

Typical curves for $n = -\frac{1}{2}$ and -2 respectively are shown in Fig. 4.1c and Fig.
4.1d.

The curves in the figure illustrate some of the many possible forms the curve
may take according to the value of n when the power law is obeyed.

It will be noticed that we have not considered the functions for negative values
of x. This is unnecessary because in biological work we very rarely need to consider negative values of x. In many ways this is fortunate because much of the
mathematics that would be involved is beyond the scope of this book.

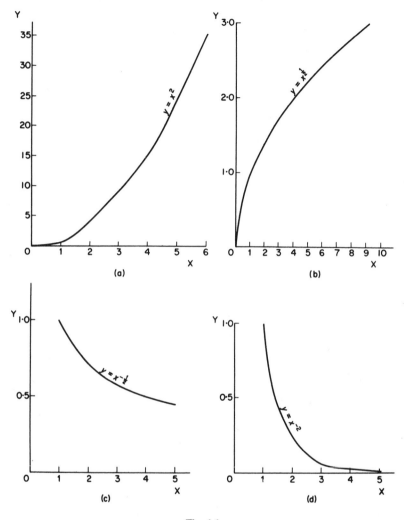

Fig. 4.1.

Example I

To determine if a power law relationship exists for the following data:

x	1	2	3	4	5	6	7
y	1.00	1.19	1.32	1.41	1.5	1.57	1.63

We draw up a table of the logarithms of the above values as follows:

x	1	2	3	4	5	6	7
$X = \log x$	0	0.3010	0.4771	0.6021	0.6990	0.7782	0.8451
y	1	1.19	1.32	1.41	1.5	1.57	1.63
$Y = \log y$	0	0.0755	0.1206	0.1492	0.1761	0.1959	0.2122

The values of $\log x$ and $\log y$ are now plotted (Fig. 4.2). It is seen that the points lie on quite a good straight line of slope $\frac{1}{4}$ which passes through the origin. If we insert these values into equation (4.3) we obtain

$$Y = \frac{1}{4}X + 0.$$

Thus $n = \frac{1}{4}$ and, since $\log A = 0$ we have $A = 1$. From these values we see that the data fit an equation of the form

$$y = x^{1/4}$$

Exercises 4.1

Test whether the following sets of data obey a power law relationship and if so determine the constants A and n:

(1)	x	1	2	3	4	5	6
	y	2.8	8.5	15.6	24.0	35.4	44.1

(2)	x	1	2	3	4	5	6
	y	2.3	1.15	0.78	0.58	0.46	0.38

(3)	x	20	30	40	50	60	70
	y	22	30.4	38.2	45.8	52.6	59.8

4.2   THE USE OF LOG-LOG PAPER

Consider the following set of data:

x	1	2	3	4	5	6	7	8	9	10
y	2.0	2.8	3.45	4.0	4.46	4.9	5.3	5.65	6.0	6.3

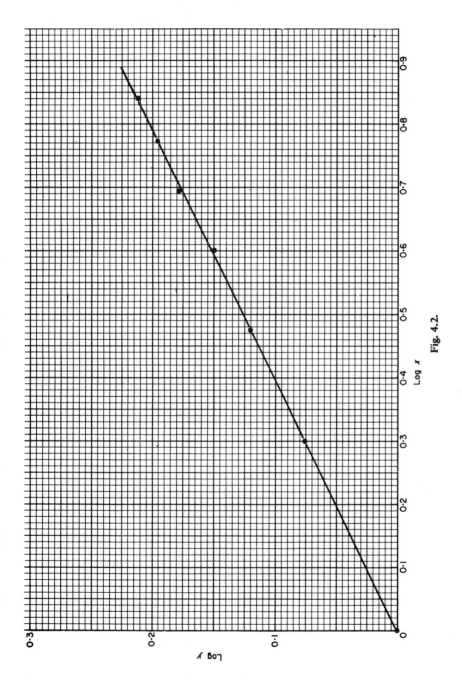

Fig. 4.2.

Clearly there is no linear relationship so we try to find out if a power law exists. We draw up the following table of logarithms of the values of x and y:

x	1	2	3	4	5	6	7	8	9	10
$\log x$	0	0.301	0.477	0.602	0.699	0.778	0.845	0.903	0.954	1.00
y	2.0	2.8	3.45	4.0	4.46	4.9	5.3	5.65	6.0	6.3
$\log y$	0.301	0.447	0.538	0.602	0.649	0.690	0.724	0.752	0.778	0.799

We draw a graph of $\log x$ against $\log y$, Fig. 4.3.

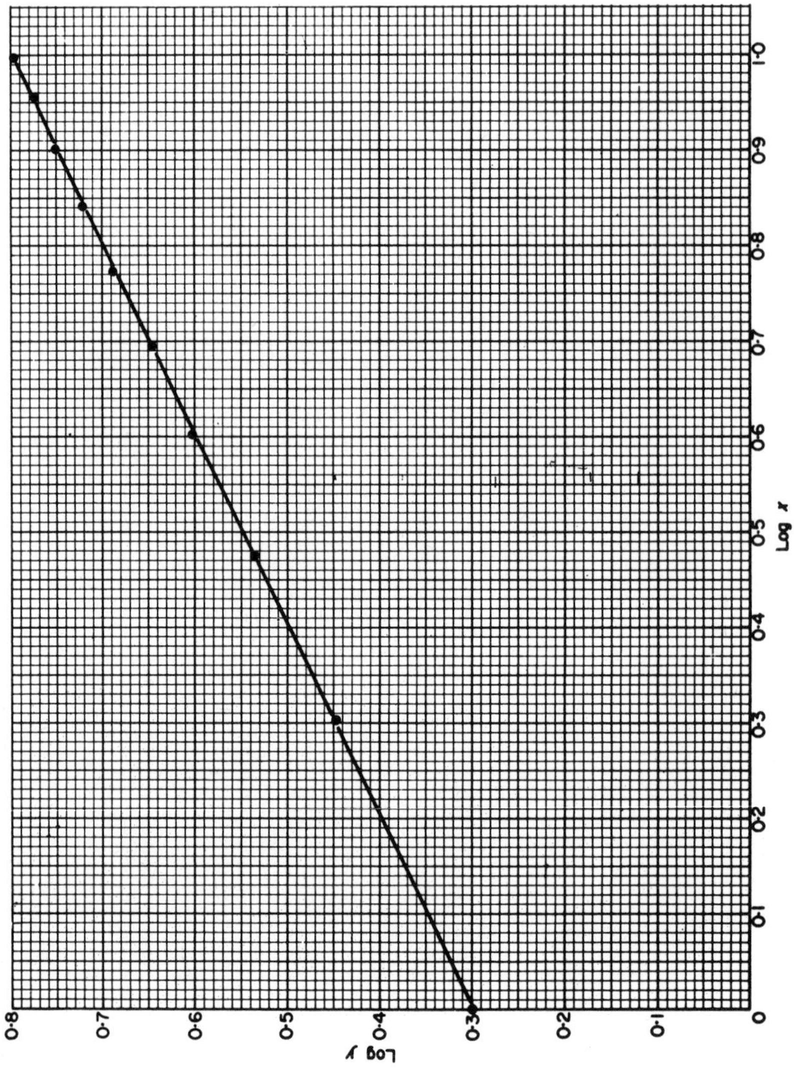

Fig. 4.3.

The slope of the line is measured and in this case it is found to be 0.5. The intercept is 0.3. The resultant equation is therefore

$$\log y = \tfrac{1}{2}(\log x) + 0.3$$
$$= \tfrac{1}{2}(\log x) + \log 2$$

or

$$y = 2x^{1/2}$$

(In this case we ignore the small difference that arises because log 2 is more accurately equal to 0.301).

A better way of plotting the data is to use a special kind of paper known as log-log paper (Fig. 4.4).

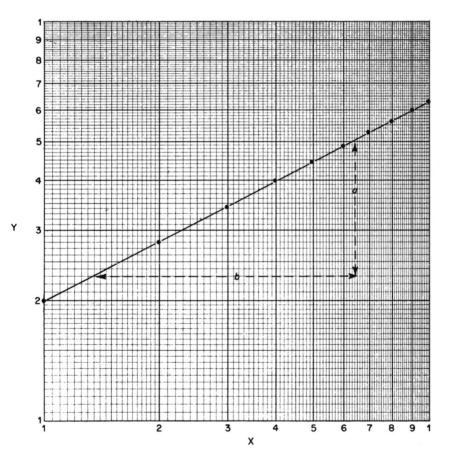

Fig. 4.4.

On log-log paper the lines are not equal distances apart but are spaced according to the logarithms of the numbers that they represent. Thus for example, the line at 9 is twice as far as the line at 3 from the bottom left hand corner because log 9 is twice the value of log 3. When we use this paper we plot the points straight on to it. There is no need to look up their logarithms.

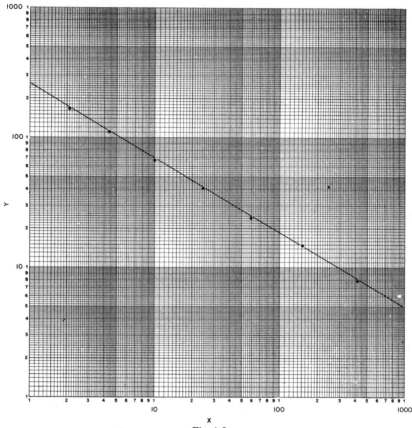

Fig. 4.5.

Our set of data lies on a straight line. The slope of this line is given by the ratio a/b as shown in Fig. 4.4. The values of a and b must be measured in absolute units and the numbers at the edge of the graph paper must be ignored. If we observe this rule we see that the ratio a/b is equal to one half. The intercept is two. Since we have plotted our data on log-log paper we write

$$\log y = \tfrac{1}{2}(\log x) + \log 2$$

or

$$y = 2x^{1/2}.$$

When we use the paper shown in Fig. 4.4 the spread of values of our data is restricted to the range 1 to 10. We could of course multiply the range by say 10 if we wanted to use data within the range 10 to 100, or we could multiply the range by any power of 10, but there is the limitation that the largest value must be no more than 10 times the smallest value in our set of data. It is possible to use other sheets of log-log paper which extend over two, three, four or more 'cycles' so that a greater range of values can be included on the same sheet.

In the graph of Fig. 4.5 data are included which have values of x from one to 1000. The values of x extend over three cycles of the paper. When using such paper it is usually convenient to mark on the axes the positions of 10, 100, 1000 as a guide when plotting the points.

Great care must be taken in the determination of the value of the intercept when setting up the equation for the points. This is particularly true if the range of the ordinate values is not the same as the range of the abscissa values.

4.3   THE RECTANGULAR HYPERBOLA

If, in the equation

$$y = Ax^n,$$

we put $n = -1$, it may be written

$$x\, y = A.$$

This equation is a special form of a family of curves known as *hyperbolas*. In particular the above equation is known as a *rectangular hyperbola*. It may be represented by a graph of the shape shown in Fig. 4.6 for $A = 1$.

It will be seen in this graph that two curves may be drawn to satisfy the equation. This is because negative values of x and y may be taken. However, as stated previously, we need only consider the curve which appears in the first quadrant or, in the case of a negative value for A, in the fourth quadrant.

In order to test whether experimental data fit a rectangular hyperbola it is not necessary to rearrange the equation by taking logarithms as in the previous section, but to write it in the form

$$y = A\, \frac{1}{x},$$

or

$$y = A\, X$$

by replacing $\frac{1}{x}$ by X. Thus, if, instead of plotting y against x, we plot y against $\frac{1}{x}$, the resultant curve becomes a straight line through the origin with slope A.

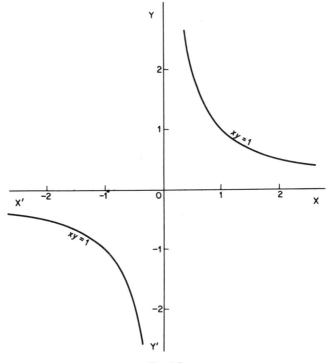

Fig. 4.6.

Example I

Determine whether the following points lie on a rectangular hyperbola:

x	0.125	0.25	0.375	0.5	0.625	0.75	0.875	1.0
y	0.97	0.48	0.32	0.25	0.19	0.16	0.14	0.12

For these data we draw up a table of values of $\frac{1}{x}$ for corresponding values of y:

$\frac{1}{x}$	8	4	2.7	2	1.6	1.3	1.14	1
y	0.97	0.48	0.32	0.25	0.19	0.16	0.14	0.12

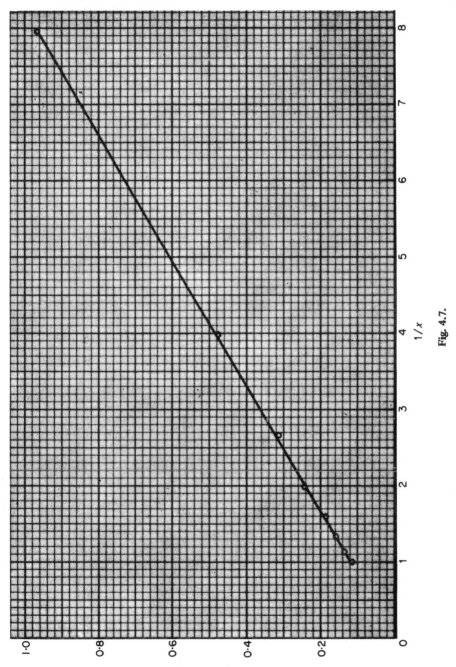

Fig. 4.7.

If these points are plotted (Fig. 4.7) it is seen that they lie on a straight line which passes through the origin and has a slope 0.12. The points thus lie on a hyperbola having an equation

$$xy = 0.12.$$

It is useful to note that in a plot such as Fig. 4.7 the points tend to cluster near the origin. If it is suspected that points lie on a hyperbola then the spacing between the higher values of x should be smaller than the spacing between the lower values of x. If this is done, then the spacing of the reciprocals will tend to be more regular.

Exercises 4.3

Test whether the following set of points lie on a rectangular hyperbola:

x	2	4	6	8	10	12	14	16
y	1.45	0.80	0.47	0.35	0.28	0.27	0.22	0.19

4.4 CHANGE OF AXES

Before we go on to consider the biological uses of the rectangular hyperbola it would be useful to consider, in general terms, how an equation is changed if the axes are shifted or rotated. In Fig. 4.8, P is a point whose co-ordinates with respect to OX and OY are (x, y). The origin is now moved to (h, k) and we want to know the co-ordinates of P with respect to the new axes $O'X'$ and $O'Y'$. Call them (x', y').

Form the geometry of the figure

$$y = PB = AB + PA = k + y'$$
$$x = OB = CO' + O'A = h + x'.$$

The suffixes are now dropped from the new co-ordinates. That is we write $h + x$ instead of x and $k + y$ instead of y.

For example, the equation

$$ax + by + C = 0$$

becomes

$$a(h + x) + b(k + y) + C = 0$$

or

$$ax + by + ah + bk + C = 0.$$

From this result it is seen that changing the origin for a straight line merely changes the intercept.

Consider now the rectangular hyperbola

$$x\,y = 6.$$

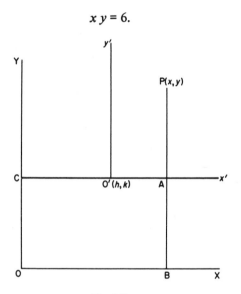

Fig. 4.8.

If we change the axes to (2, 3) we write

$$(x + 2)\,(y + 3) = 6$$

$$xy + 2y + 3x = 0.$$

Now let us consider what happens to an equation if we rotate the axes. P (Fig. 4.9) is the point (x,y) with respect to the axes OX, OY.

The axes are rotated through an angle θ and we want to know the new co-ordinates of P which we will call (x',y'). In the figure perpendiculars are dropped from R to meet PQ in S and OX in T. From the geometry of the resultant figure we may write

$$PR = y',$$
$$OR = x',$$
$$PQ = y,$$
$$OQ = x.$$

Now since

$$\hat{QPR} = \hat{ROQ} = \theta,$$

then

$$
\begin{aligned}
y = PQ &= PS + SQ = PS + RT \\
&= PR \cos \theta + OR \sin \theta \\
&= y'\cos \theta + x' \sin \theta, \\
x = OQ &= OT - QT = OT - SR \\
&= OR \cos \theta - PR \sin \theta \\
&= x'\cos \theta - y'\sin \theta.
\end{aligned}
$$

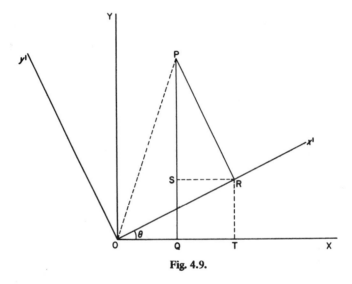

Fig. 4.9.

When we want to rotate the axes through an angle θ we substitute

$$
\begin{aligned}
x &= x \cos \theta - y \sin \theta, \\
y &= x \sin \theta + y \cos \theta.
\end{aligned}
$$

For example the equation

$$ax + by + C = 0$$

becomes

$$a(x \cos \theta - y \sin \theta) + b(x \sin \theta + y \cos \theta) + C = 0$$

or

$$(a \cos \theta + b \sin \theta) x + (b \cos \theta - a \sin \theta) y + C = 0.$$

If it is required both to change the position of the origin and to rotate the axes the two operations are performed in turn.

Exercises 4.4

(1) Transform to parallel axes through the point $(4, -5)$ the equations

    (a) $y^2 - 4x + 10 = 0$
    (b) $y = \cos(x - 4) + \sin(x + 5)$
    (c) $x^2 = (y - 4)^2$

(2) Transform to axes inclined at $45°$ to the original axes the equations

    (a) $x^2 - y^2 = 1$
    (b) $y = x + 1$
    (c) $xy = 2$.

4.5   THE MICHAELIS-MENTON RELATIONSHIP

The rate at which an enzyme-catalysed reaction proceeds is dependent upon the concentrations of the enzyme and the substrate. If we have a fixed amount of enzyme and vary the amount of substrate, it is seen that the initial velocity of the reaction does not increase linearly with the substrate concentration but that it rises sharply at the low concentration, then the slope decreases until at the higher substrate concentrations a limit to the initial velocity of the reaction is reached.

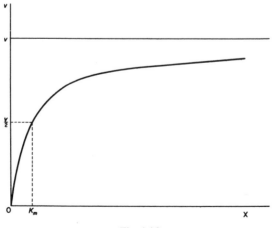

Fig. 4.10.

This curve is produced empirically during an experiment but a precise equation has been worked out and this equation is known as the Michaelis-Menton or Briggs-Haldane equation and it takes the form

$$v = \frac{Vx}{x + Km} \qquad (4.4)$$

where v is the initial rate of the reaction, x is the substrate concentration, V is the maximum value of the initial velocity and K_m is known as the Michaelis constant and has a characteristic value for each reaction.

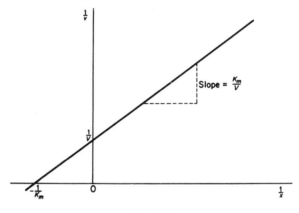

Fig. 4.11.

If, in equation (4.4), we put

$$v = \frac{V}{2}$$

then $x = K_m$. Thus the Michaelis constant corresponds to that substrate concentration at which half the limiting value of the velocity is developed.

Although equation (4.4) describes a rectangular hyperbola (Fig. 4.10) (and this may be proved by making a suitable choice of new axes to bring the equation to the form $v. x =$ constant) we do not bring it to a linear form by the method suggested in Section 4.3. because there are two constants in this equation K_m and V. In order to determine these quantities from experimental data we usually transpose the equation to the form

$$\frac{1}{V} = \frac{x + K_m}{Vx} = \frac{1}{V} + \frac{K_m}{Vx} \qquad (4.5)$$

If now we plot $1/v$ against $1/x$ we should obtain a straight line of slope K_m/V and intercept $1/V$ (Fig. 4.11). The equation in this form makes the determination of the constants much simpler but there is the disadvantage that, because we take reciprocal values for x and v, the points on the line tend to become bunched together nearer to the origin and progressively wider apart on the parts of the line further from the origin.

The graph shown in Fig. 4.11 is known as a 'Lineweaver-Burk' plot and an estimate of the Michaelis constant can be obtained simply by extending the line to cross the $1/x$ axis.

4.6   FORCE-VELOCITY CURVE FOR MUSCLE

It is a matter of common experience that, when a limb moves to raise a load, the actual speed at which the load is raised depends upon the magnitude of the load. Small loads can be lifted at a greater speed than heavy loads, some loads are too heavy to be lifted at all but at the other extreme there is a limiting speed to the rate at which the smallest loads can be lifted. If an isolated muscle is studied by suspending loads on it and then stimulating the muscle to contract, a relationship between the load and the speed of contraction can be obtained. According to A. V. Hill an empirical relationship between the load (P) acting upon the muscle and the speed of shortening (V) is given by

$$(P + a) V = (P_0 - P) b \qquad (4.6)$$

where a, b and P_0 are constants which are characteristic for the particular species and type of muscle. It seems that all the muscles so far tested obey this relationship but have differing values for the constants. A typical curve for the force-velocity relationship of frog sartorious muscle, as given by A. V. Hill[†], is reproduced in Fig. 4.12.

The constant P_0 represents the maximum force that the muscle can exert in an isometric contraction, i.e. where the muscle is not allowed to shorten. This is shown by putting $V = 0$ in equation (4.6) to obtain $P = P_0$.

We must now try and rearrange the equation into a linear form so that the other two constants may be obtained.

First of all let us rearrange the equation as follows:

$$
\begin{aligned}
(P + a)(V + b) &= (P_0 - P)b + b(P + a) \\
&= (P_0 + a)b.
\end{aligned}
\qquad (4.7)
$$

† (A. V. Hill (1938). The Heat of Shortening and the Dynamic Constants of Muscle, *Proc. R. Soc.* B, **126**, 136-195.

The term on the right hand side of this equation is a constant. In fact the equation is a hyperbola which, if the values of a and b were known, could be put in the form

$$PV = \text{constant}$$

merely by changing the position of the axes.

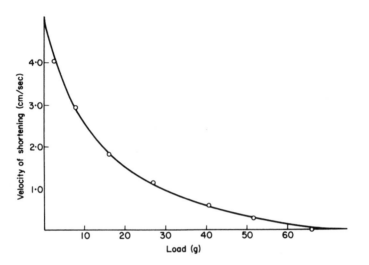

Fig. 4.12. Force-velocity relation, in after-loaded isotonic contractions of a frog sartorius at $0°C$. Tetanus, 11.4 shocks/sec. Initial load 2.5 g. Muscle 0.165 g, $l_0 = 3.8$ cm. Each circle in the figure is the mean of two observations in a series and reverse.

At first sight it might seem that the equation either in the form (4.6) or in the form (4.7) is not amenable to a transposition to linear form. However, if we take (4.6) and put it in the form

$$PV + aV = (P_0 - P)b$$

and then divide throughout by V we obtain:

$$P + a = b(P_0 - P)/V.$$

So that, if we write

$$Y = P$$

and

$$X = \frac{P_0 - P}{V}$$

we obtain

$$Y + a = bX.$$

This brings the relationship to linear form, the values of X and Y may be obtained for each experimental determination of V and P, and hence the values of a and b are readily obtainable from the resultant straight line.

4.7   THE EXPONENTIAL LAW

An equation to the form

$$y = A \cdot \exp[mx] \qquad (4.8)$$

where A and m are constants is often encountered in biological work. The curves of $y = \exp[mx]$ (i.e. for $A = 1$) for sample values of m are shown in Fig. 4.13.

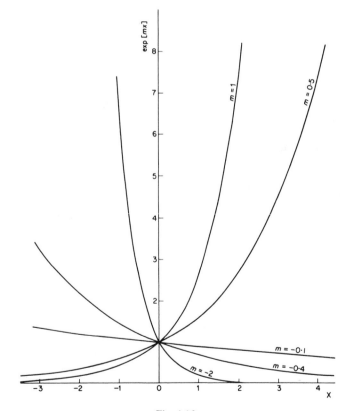

Fig. 4.13.

It is seen that whatever the value of m, whether it is positive or negative, the curve always goes through the point $(0,1)$. Also the curve approaches the X-axis but never crosses it. In fact it tends to become a tangent to the curve, i.e. the X-axis is an asymptote of the curve $y = \exp[mx]$.

If A takes a value which is not unity then the curve cuts the Y-axis at the point $(0,A)$. The X-axis is an asymptote to the curve $y = A\exp[mx]$ whatever the value of A.

In order to test if a set of data obeys the exponential law we take natural logarithms of each side of the equation so that

$$y = A \,.\, \exp[mx]$$

becomes

$$\ln y = \ln A + mx. \tag{4.9}$$

Since $\ln A$ is a constant we can plot $Y (= \ln y)$ against x and obtain a straight line whose equation is

$$Y = \ln A + mx.$$

The slope of this straight line is m and from $\ln A$, the intercept on the Y-axis, the value of A is found.

Example

The concentrations of a substance within the circulation at various times are given in the following table:

Time (min)	0	10	20	30	40	50	60
Concentration (μg per c.c.)	20.1	10.3	6.5	4.7	2.6	1.8	1.2

To determine whether the removal of the substance follows an exponential law, we first of all plot the points as in Fig. 4.14a. The curve has the appearance of an exponential so we now apply a further test to see if the points lie on a curve of the form

$$C = A \exp[kt]$$

where C is the concentration and t the time. By taking natural logarithms of the concentrations we draw up the following table:

Time	0	10	20	30	40	50	60
ln C	3.0007	2.3321	1.8718	1.5476	0.9555	0.5878	0.1823

When these points are plotted as in Fig. 4.14b it is seen that they fall on a reasonably straight line so we may assume that the points lie on the line

$$\ln C = \ln A + kt$$

and from the intercept we see that

$$\ln A = 2.84$$

so that

$$A = 17.1,$$

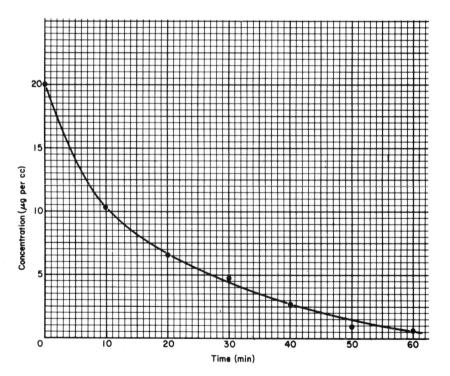

4.14a

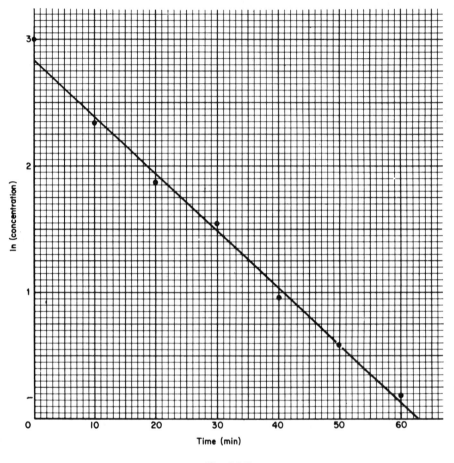

Fig. 4.14b.

and from the slope of the curve we see that

$$k = -\frac{2.84}{63}$$

$$= -0.045$$

Thus the data seem to indicate that the substance is removed from the circulation according to an exponential law having the form

$$C = 17.1 \exp[-0.45t]$$

or we put it in the form

$$C = 17.1 \exp[-t/22.2]$$

where the time constant = 22.2 min.

In this example we have not used either of the methods for determining the line of best fit because it is seen that in Fig. 4.14b the points lie on a straight line reasonably well.

Exercises 4.7

The following sets of data follow the relationship

$$y = A \exp(Bx).$$

Determine the constants A and B.

(a)

x	0.5	1.5	2.5	3.5	4.5
y	33.5	43.0	55.0	70.0	89.5

(b)

x	1.0	2	3	4	5	6
y	5.6	4.0	2.8	2.01	1.42	1.02

4.8 SEMI-LOG PAPER

The data of the example from the previous section are now considered again. This time, instead of using ordinary graph paper, we use a special type of graph paper in which the divisions along the X-axis are evenly spaced but the divisions along the Y-axis are spaced logarithmically. Consider the graph of Fig. 4.15. It is seen that the sheet of graph paper extends over two cycles so that we may use it to plot data over the range from 1 to 100 or from 0.1 to 10 etc. In our particular case the data extend from 1.2 to 20.1 so we use the 1 to 100 range and mark the graph paper accordingly on the left hand margin. The time scale is marked off in the normal way. We do not plot our data straight onto the paper without taking logarithms of the concentration. It is seen that the points are now arranged in exactly the same way as when the graph of Fig. 4.14b was plotted. The value for

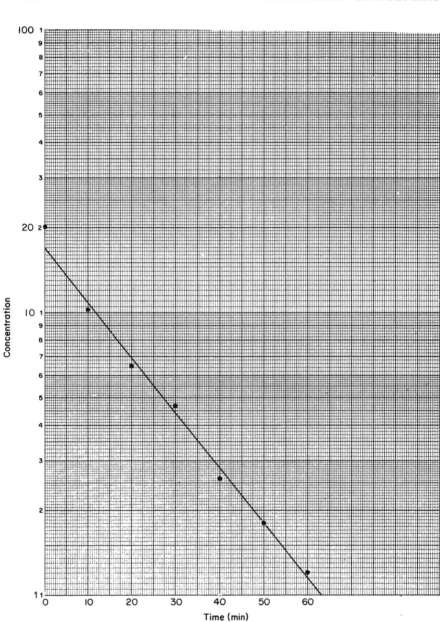

Fig. 4.15.

A is read straight off at the intercept of the line with the Y-axis. The calculation of the slope of the line needs care. It is best done by taking the logarithm of the intercept on the Y-axis and dividing by the intercept on the X-axis.

The use of this paper enables graphs containing many points to be quickly plotted without having to determine the logarithms of their ordinates beforehand.

4.9 FUNCTIONS OF THE FORM $y = A - B\exp[-kt]$

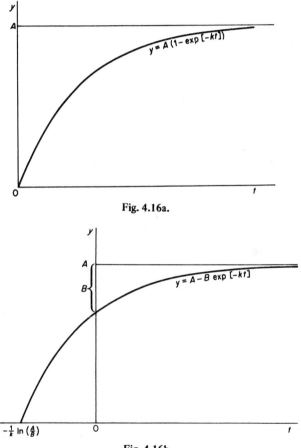

Fig. 4.16a.

Fig. 4.16b.

Often one comes across situations in which the observed variable seems to be changing exponentially with time but that, instead of the simple form considered in the previous section, the relationship takes the form quoted above. In this case there are three constants to be determined and usually they are positive. Quite often the equation is simplified so that $A = B$ and then takes the form

$$y = A(1 - \exp[-kt]).\qquad(4.10)$$

The shape of this curve is indicated in Fig. 4.16a.

The curve goes through the origin and is asymptotic to the line $y = A$.

If on the other hand the equation is

$$y = A - B\exp[-kt],\qquad(4.11)$$

then a typical curve has the form shown in Fig. 4.16b. In such a curve when $t = 0$ then $y = A - B$; when t tends to large values y tends to the value A; when $y = 0$, $\exp[-kt] = A/B$ so that

$$t = -(1/k) \ln (A/B).$$

If B is greater than A, the curve cuts the t-axis at a positive point and cuts the y-axis at a negative point.

Let us consider equation (4.10). Although there are only two constants A and k to be determined, it is not possible to do this simply by rearranging the equation. If we take natural logarithms of the equation, we could only write

$$\ln y = \ln A + \ln (1 - \exp [-kt])$$

and it is true that this could be a linear equation if we were to write

$$Y = \ln y,$$
$$X = \ln (1 - \exp[-kt])$$

But this does not help us because without a knowledge of the value of k we cannot calculate the values of $\ln (1 - \exp[-kt])$. In fact, the determination of the two constants in equation 4.10 is no easier than determining the values of the constants A, B and k of equation (4.11). The equations may be rewritten in the form

$$(A - y) = \exp[-kt]$$
$$(A - y) = B.\exp[-kt]$$

By taking logarithms we obtain

$$\ln (A - y) = -kt$$

and

$$\ln (A - y) = \ln B - kt.$$

At first sight it would appear that we are no nearer the solution than previously because we do not know the value of A and therefore cannot calculate the values of $\ln (A - y)$. However, if we look at Fig. 4.16 we see that A is the asymptote— no values of y exceed A and, as t increases, the values of y should become closer to the values of A. Therefore, if we plot the points, we should be able to make an estimate of the value of A from the trend of the curve. This may not always be easy to do because either the points may not give a smooth curve or because the points may not lie on that part of the curve in which y has values near to the asymptote. In such cases one must usually make several estimations of the value of A and go through the process of plotting values of $\ln (A - y)$ against t for each estimated value of A and then to decide which plot gives the best linear fit.

Example

In the following table the first two columns give the measurements of a quantity at various times during an experiment. These are plotted on the curve of Fig. 4.17

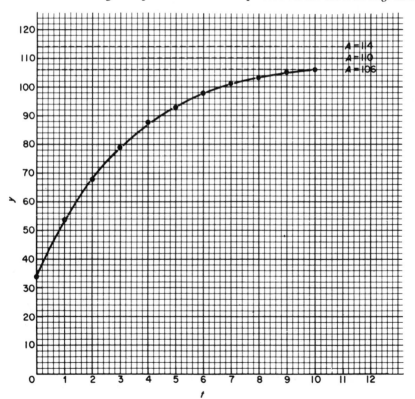

Fig. 4.17.

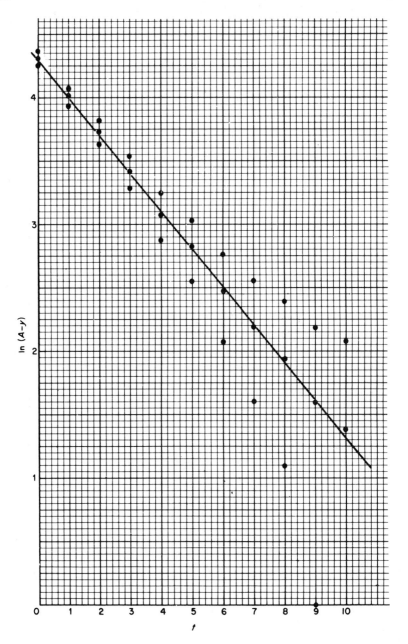

Fig. 4.18.

and estimations of A are made. The other columns give values of $(A - y)$ and $\ln (A - y)$ for each estimation of A.

		$A = 106$		$A = 110$		$A = 114$	
t	y	$A - y$	$\ln(A - y)$	$A - y$	$\ln(A - y)$	$A - y$	$\ln(A - y)$
0	34	72	4.28	76	4.33	80	4.38
1	54	52	3.95	56	4.03	60	4.09
2	68	38	3.64	42	3.74	46	3.83
3	79	27	3.30	31	3.43	35	3.55
4	88	18	2.89	22	3.09	26	3.26
5	93	13	2.56	17	2.83	21	3.04
6	98	8	2.08	12	2.48	16	2.77
7	101	5	1.61	9	2.20	13	2.56
8	103	3	1.10	7	1.95	11	2.40
9	105	1	0	5	1.61	9	2.20
10	106	0	$-\infty$	4	1.39	8	2.08

The values of A have been estimated from the graph of Fig. 4.17. The first estimate $A = 106$ would seem to be too low because the last point on the curve has this value and it suggests that the trend is for it to go slightly higher if further points were available. However, this might be a good point at which to start and the curves of $\ln(A - y)$ for the three values of A are given in Fig. 4.18.

It may readily be seen that the value $A = 110$ gives the straightest line. The linearity is such that the slope may be estimated with a reasonable degree of accuracy without doing a best fit calculation and thus we obtain the values

$$k = 0.293; \ln B = 4.3 \text{ or } B \simeq 74.$$

The points given in the example are thus seen to obey the relationship

$$y = 110 - 74\exp[-0.293t]$$

5

DIFFERENTIATION

5.1 INTRODUCTION

In this chapter we move on from the elementary algebraic operations that we have used so far and enter into that branch of higher mathematics known as the calculus. This is usually subdivided into two branches, the differential calculus and the integral calculus.

Both branches of the calculus nowadays have increasingly important applications in certain branches of biology, and therefore biologists must equip themselves with a knowledge and facility of the subject both from the point of view of their research and to enable them to understand the literature.

The differential calculus involves the calculation of the rate of change of a function with respect to its independent variable. In order to do these calculations we must introduce certain mathematical concepts and make ourselves familiar with their meaning. Many examples are given for the student to work out for himself. It must be stressed that the working out of as many examples as possible is most important if one wants to thoroughly master the subject.

5.2 GRADIENTS

We are familiar from the previous chapters with the fact that in the equation of the straight line

$$y = mx + C$$

the constant m represents the slope of the line and it is the tangent of the angle that the line makes with the X-axis. The constant m is also known as the gradient of the straight line and, of course, in this case its value remains unchanged.

Now let us consider the case of a non-linear function of which Fig. 5.1 is an example.

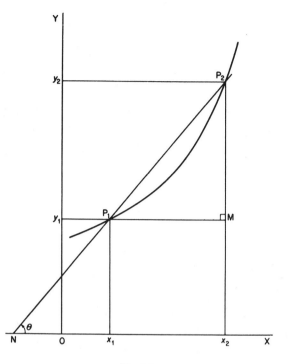

Fig. 5.1.

When x has the values x_1 and x_2, corresponding values of y are y_1 and y_2. Let P_1 and P_2 be points on the curve which have co-ordinates (x_1, y_1) and (x_2, y_2) respectively. The line $P_1 M$ is drawn by taking a perpendicular from P_1 to the ordinate through P_2. Finally, we join the points P_1, P_2 by a straight line which is continued to cut the X-axis at the point N and the line $P_1 P_2$ makes an angle with the X-axis equal to θ.

The difference in values of the ordinates of the points P_2 and P_1 is $y_2 - y_1$, while the difference in their abscissae is $x_2 - x_1$. The ratio

$$\frac{y_2 - y_1}{x_2 - x_1}$$

gives a measure of the mean rate of change of y with x over the range of values of x from x_1 to x_2. If now we consider Fig. 5.1 it is seen that $y_2 - y_1 = P_2 M$ and $x_2 - x_1 = P_1 M$ so that

$$\frac{y_2 - y_1}{x_2 - x_1} = \frac{P_2 M}{P_1 M},$$

and since the triangle $P_2 M P_1$ has a right angle we can write

$$\frac{P_2 M}{P_1 M} = \tan \theta.$$

We can thus say that the mean rate of change of y with x over the range of values x_1 to x_2 is given by the slope of the line through $P_1 P_2$ with the X-axis.

As an example consider the curve

$$y = x^3 + 2.$$

When $x = 1, y = 3$ and when $x = 2, y = 10$.

The average rate of change of y with x over the range of values $x = 1$ and $x = 2$ is given by

$$\frac{y_2 - y_1}{x_2 - x_1} = \frac{10 - 3}{2 - 1} = 7$$

So far we have not made any restriction upon the magnitude of the value of $x_2 - x_1$. Let us now consider what happens when $x_2 - x_1$ becomes very small or even tends to zero. To do this we keep x_1 at the value 1 and alter the value of x_2 so that $x_2 - x_1$ changes in value.

If $x_2 = 1.1$, then $y_2 = 3.331$ so that

$$\frac{y_2 - y_1}{x_2 - x_1} = \frac{0.331}{0.1} = 3.31.$$

If $x_2 = 1.01$ then $y_2 = 3.0303$ (to five significant figures) so that

$$\frac{y_2 - y_1}{x_2 - x_1} = \frac{0.0303}{0.01} = 3.03.$$

If $x_2 = 1.001$, we can show that

$$\frac{y_2 - y_1}{x_2 - x_1} \text{ is approximately 3.003.}$$

If we continue our calculation we can show that as $x_2 - x_1$ gets smaller and smaller then the value of

$$\frac{y_2 - y_1}{x_2 - x_1}$$

gets nearer and nearer to the value 3. Also it is noted that as the two points get closer and closer together the straight line which joins them becomes the *tangent* to the curve. We may thus reasonably state that, for the chosen example, the slope of the tangent to the curve

$$y = x^3 + 2$$

for $x = 1$ is equal to 3. By similar reasoning it may be shown that for $x = 2$ the slope of the tangent to the curve is equal to 14.

Let us now consider the above results in a slightly different way. We still use the point $(1, 3)$ on the curve for (x_1, y_1) but choose the value of x_2 to have the value $1 + h$ where h is any number. The corresponding value of y_2 will be

$$(1 + h)^3 + 2,$$

and the quantity

$$\frac{y_2 - y_1}{x_2 - x_1}$$

will have the value

$$\frac{(1 + h)^3 - 1}{h}$$
$$= \frac{1 + 3h + 3h^2 + h^3 - 1}{h}$$
$$= 3 + 3h + h^2.$$

As h becomes smaller and smaller the value of

$$\frac{y_2 - y_1}{x_2 - x_1}$$

approaches the value 3.

Note that so far we have considered only tangents which make acute angles with the X-axis and in such cases we say that y increases with increasing x. If the tangent makes an angle greater than $90°$ its tangent is negative and from Fig. 5.2 it is seen that y decreases as x increases.

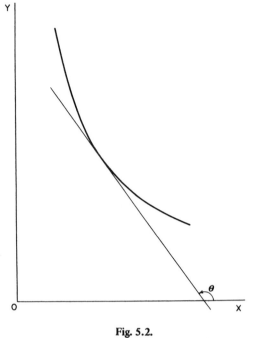

Fig. 5.2.

Exercises 5.2

(1) By methods similar to those explained above, determine the slope of the tangent to the curve

$$y = x^2 + 2x$$

at values of $x = 0.5, 3$ and 10.

(2) What is the slope of the tangent to the curve

$$y = (x + 2)^2$$

when $x = 0, 1$ and 2?

(3) What is the slope of the tangent to the curve

$$y = x^2 - 2x$$

when $x = 1, 1.5$ and 0.8?

5.3   RATES

Suppose that we observe the growth of a plant over a period of time. At various intervals we measure the height of the plant and make a graph of height against time as in the figure. In this case we do not get a straight line graph.

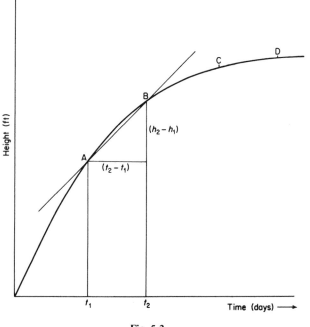

Fig. 5.3.

Consider the two points A and B on the curve. Let their co-ordinates be (t_1, h_1) and (t_2, h_2). These two points indicate that the plant has grown a height $(h_2 - h_1)$ ft over a period of $(t_1 - t_2)$ days. The average rate of growth over this period is thus

$$\frac{(h_2 - h_1)}{(t_2 - t_1)} \text{ ft per day.}$$

Note that the units of the mean rate of growth are given as ft per day. It is very important always to express the rate of growth in the correct units. If the height of the plant had been measured in metres and time in weeks then the rate of growth would be expressed in metres per week. It is of course possible to convert the units ft per day into metres per week if the appropriate conversion factors are known. When a rate is involved, the denominator is always in units of time but the numerator may have any units. For instance, the rate of growth of a population may be expressed as people per year, a chemical reaction may proceed so that a substance is broken down at a rate of so many grammes a second, an animal may respire to consume so many litres of oxygen per second, a nerve cell may fire off action potentials at a rate of so many action potentials per second. These are examples of rates, i.e. quantities whose variations with respect to time are measured.

In the example used in Fig. 5.3 we considered the average rate of growth of the plant over a period of time, and it is clear from the shape of the curve that the mean rate of growth when the two points A and B are considered is not the same as the mean rate of growth when the two points C and D are considered. This is obvious for the extreme case when the plant ceases to grow for beyond this point the rate of growth will be zero.

The rate of growth at any point in time is obtained by making the point B on the line so close to the point A that the line AB becomes a tangent to the curve and the slope of this line (i.e. the mean rate of growth over the period $t_2 - t_1$) is in fact the rate of growth at time t_1.

Consider as an example a sense organ which receives a stimulus which causes action potentials to be produced. At a particular time t sec. from the start of the stimulus the total number of action potentials $N(t)$ might be given by the formula

$$N(t) = 6t + \frac{1}{t^2 + 1} - 1.$$

We can determine the total number of action potentials produced after a total time $t + \tau$ by substituting $(t + \tau)$ instead of t in the formula. Thus

$$N(t + \tau) = 6(t + \tau) + \frac{1}{(t + \tau)^2 + 1} - 1.$$

The mean rate of production of action potentials over the interval τ is thus

$$\frac{N(t + \tau) - N(t)}{(t + \tau) - t} = \frac{1}{\tau}\left\{6\tau + \frac{t^2 + 1 - (t + \tau)^2 - 1}{\{(t + \tau)^2 + 1\}\{t^2 + 1\}}\right\}$$

$$= \frac{1}{\tau}\left\{6\tau - \frac{2t\tau + \tau^2}{\{(t + \tau)^2 + 1\}\{t^2 + 1\}}\right\}$$

$$= 6 - \frac{2t + \tau}{\{(t + \tau)^2 + 1\}\{t^2 + 1\}}$$

action potentials per second.

As the interval gets smaller and smaller, i.e. as τ gets nearer to zero, the mean rate of production of action potentials over this small interval becomes more nearly equal to the rate of production of action potentials at time t. Thus, in our example, as τ approaches the value zero the mean rate of production of action potentials approaches the value

$$6 - \frac{2t}{(t^2 + 1)^2}$$

action potentials per second.

We have thus derived a formula which enables us to calculate the rate of production of action potentials at *any point in time* after the onset of the stimulus. The rate of production at a particular point in time is merely obtained by substituting the appropriate value of t into the formula. Thus the rate of production after 4 secs. from the onset of the stimulus is given by

$$6 - \frac{2 \times 4}{(4^2 + 1)^2}$$

$$= 6 - \frac{8}{(17)^2}$$

$= 5.97$ action potentials per sec. to three significant figures.

Exercises 5.3

(1) At the start of an experiment a bacterial culture was found to contain 10^5 individuals. The growth of the population was observed and it was found that at any subsequent time t from the start of the experiment, the number of individuals $N(t)$ could be expressed according to the formula

$$N(t) = 10^5(1 + 2t + t^2)$$

If t is expressed in hours, find a formula for the rate of growth of the population at any time t, and in particular calculate the rate of growth for $t = \frac{1}{2}$ hour and $t = 2$ hours.

5.4   LIMITS

In the previous two sections we saw how it was possible to calculate the gradient of the tangent to a curve at a particular point by first of all calculating the slope

of a line through two points on the curve and then bringing these points closer
and closer together until they become virtually coincident. In this process we found
the *limiting* value of the slope of the line through two points as they approached
each other on the curve. Since this idea of taking limits or limiting values is fundamental to the calculus, this section is devoted to a more precise, but obviously
very incomplete, discussion of the nature of a limit.

Suppose y is a function of x. If x takes a series of values which gradually
approach a fixed number a, then the corresponding values of y will generally
approach a fixed number b, and y may be made as near to the value of b as we
please by making x near enough to a. In such a case b is said to be the limit of y
as x approaches a. In mathematical shorthand this may be written

$$\underset{x \to a}{\text{Lt}}\ y = b.$$

Sometimes in books we might find the above expression in the form

$$\underset{x = a}{\text{Lt}}\ y = b.$$

The former version is perhaps preferable because in many cases x cannot be
made equal to a but only as near to it as we please without being exactly equal to
it. This is true in the case of the function

$$f(x) = \frac{x^2 - 1}{x - 1}$$

The value of this function is obtained by direct substitution of any value of
x *except $x = 1$.* We can obtain the values

$$f(0) = 1,$$
$$f(2) = 3,$$
$$f(\tfrac{1}{2}) = \tfrac{3}{2}, \text{ etc.,}$$

but when we try to calculate $f(1)$ the numerator and the denominator of the
function are both zero so that the function is indeterminate. Instead we must
take a series of values of x which approach 1 but differ from it by a small amount.
For example

$$f(0.9) = \frac{(0.9)^2 - 1}{0.9 - 1} = 1.9$$

$$f(0.99) = 1.99$$

$$f(0.999) = 1.999.$$

Likewise, we may take values of x slightly greater than 1,

$$f(1.1) \quad = 2.1$$
$$f(1.01) \quad = 2.01$$
$$f(1.001) = 2.001.$$

From both sets of values it is seen that values of $f(x)$ can be made to differ from 2 by as small a quantity as we wish by taking x sufficiently close to unity. Hence we may write

$$\underset{x \to 1}{\mathrm{Lt}} \left\{ \frac{x^2 - 1}{x - 1} \right\} = 2.$$

In calculating the values of the function a short cut may be used by putting

$$\frac{x^2 - 1}{x - 1} = \frac{(x - 1)(x + 1)}{(x - 1)} = (x + 1).$$

This method cannot correctly be used for the case where $x = 1$ because then the numerator and the denominator are both zero. Division by zero is not permissible. It is seen, however, that $x + 1$ does in fact take the value 2 when we put $x = 1$, so here we encounter an example where incorrect reasoning may lead to the correct answer.

The example that we have used gives a good illustration of the case where a fraction tends to a finite value when both its numerator and denominator tend to zero. This is a type of limit which is very common in the differential calculus; it will be seen in the following sections that differential coefficients are limits of fractions whose numerators and denominators both tend to zero.

Exercises 5.4

Evaluate by the method explained above the following limits:

(1) $\underset{x \to 2}{\mathrm{Lt}} \left\{ \dfrac{x^2 - 4}{x - 2} \right\}$

(2) $\underset{x \to 3}{\mathrm{Lt}} \left\{ \dfrac{x^3 - 27}{x - 3} \right\}$

(3) $\underset{x \to 2}{\mathrm{Lt}} \left\{ \dfrac{x^3 - 8}{x^2 - 4} \right\}$

(4) $\underset{x \to -1}{\mathrm{Lt}} \left\{ \dfrac{x^3 + 1}{x^2 - 1} \right\}$

(5) $\underset{x \to 2}{\mathrm{Lt}} \left\{ \dfrac{x^5 - 32}{x^3 - 8} \right\}$

(6) $\underset{x \to b}{\mathrm{Lt}} \left\{ \dfrac{x^3 - b^3}{x - b} \right\}$

(7) $\underset{x \to 3}{\mathrm{Lt}} \left\{ \dfrac{x^2 - 9}{x - 3} \right\}$

(8) $\underset{x \to 0}{\mathrm{Lt}} \left\{ \dfrac{x^2 - 2x}{x} \right\}$

(9) $\underset{x \to 2}{\mathrm{Lt}} \left\{ \dfrac{x^2 - x - 2}{x(x - 2)} \right\}$

(10) $\underset{x \to -1}{\mathrm{Lt}} \left\{ \dfrac{(x - 1)^2 - 4}{x + 1} \right\}$.

5.5   THE DIFFERENTIAL COEFFICIENT

In previous sections we showed how to find the gradient to a curve at a particular point. In the example we chose this process was quite simple. In many mathematical problems the calculation of the gradient by such a method would be very tedious but by means of the differential calculus we can reduce the labour to a minimum.

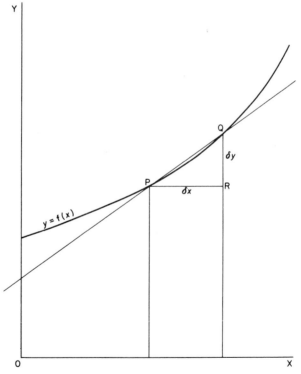

Fig. 5.4.

Suppose we have the relation

$$y = f(x)$$

and we want to find a formula for the gradient of this curve for any value of x. Previously we chose two points on the curve whose abscissae differed by a quantity h. In higher mathematics we use a special symbol for small values of h. This symbol is δx and is called 'delta x'.

The meaning of δx is best explained by reference to Fig. 5.4.

This shows a portion of the graph of $y = f(x)$. We consider any point $P(x, y)$ on this curve. Let Q be a neighbouring point on the curve. Its abscissa will differ from the abscissa of the point P by a small amount. Thus, if the abscissa of P is x, then the abscissa of Q is $x + \delta x$ where δx denotes a small increase or increment in the value of the independent variable x. Now let us consider the ordinates of the two points P and Q. The ordinate of P is y. The symbol δy is used to denote the change in the value of y produced by a change δx in the value of x. Thus, if the co-ordinates of P are (x, y), then the co-ordinates of Q are $(x + \delta x, y + \delta y)$ and both points lie on the curve $y = f(x)$. In the figure we draw in the ordinates at P and Q and draw PR parallel to the X-axis.

From this figure it is easily seen that, in the triangle PQR, PR $= \delta x$ and QR $= \delta y$. The slope of the chord PQ is given by QR/PR $= \delta y/\delta x$. The gradient of the curve at P is given by the limiting value of the ratio $\delta y/\delta x$ as δx tends to zero. Also, from the figure it is seen that

$$QR = \delta y = (y + \delta y) - y$$
$$= f(x + \delta x) - f(x),$$

and

$$\frac{\delta y}{\delta x} = \frac{f(x + \delta x) - f(x)}{\delta x}. \tag{5.1}$$

The gradient of the curve at the point P is given by

$$\underset{\delta x \to 0}{\mathrm{Lt}} \left(\frac{\delta y}{\delta x} \right) = \underset{\delta x \to 0}{\mathrm{Lt}} \left\{ \frac{f(x + \delta x) - f(x)}{\delta x} \right\}. \tag{5.2}$$

This relationship is very important; it forms the basis of the differential calculus.

The quantity

$$\frac{dy}{dx} = \underset{\delta x \to 0}{\mathrm{Lt}} \left(\frac{\delta y}{\delta x} \right)$$

is known as the *differential coefficient* or the *derivative* of y with respect to x.

Several notations are in common use for the differential coefficient:

$$\frac{dy}{dx}, \ f'(x), \ Dy, \ D_x y, \ y', \ \frac{d}{dx} f(x).$$

The process of finding the derivative is called *differentiation* with respect to x. Let us now illustrate what we have said so far by considering the function

$$y = x^3 + 2. \tag{5.3}$$

Let us choose any point P on the curve with co-ordinates (x, y). A neighbouring point Q on the curve will have coordinates $(x + \delta x, \ y + \delta y)$ and the relationship

$$y + \delta y = (x + \delta x)^3 + 2 \tag{5.4}$$

must therefore be true.

We subtract equation (5.3) from equation (5.4) to obtain

$$\begin{aligned} \delta y &= (x + \delta x)^3 - x^3 \\ &= 3x^2 \, \delta x + 3x(\delta x)^2 + (\delta x)^3. \end{aligned} \tag{5.5}$$

We can now write

$$\frac{\delta y}{\delta x} = \frac{3x^2 \, \delta x + 3x(\delta x)^2 + (\delta x)^3}{\delta x} \tag{5.6}$$

It is seen from equation (5.5) that, as δx approaches zero, δy tends to zero. However, if δx is not zero, we may write equation (5.6) as follows:

$$\frac{\delta y}{\delta x} = 3x^2 + 3x \, \delta x + (\delta x)^2.$$

As δx approaches zero then $\delta y / \delta x$ approaches the value $3x^2$. Thus, for the example chosen, the limiting value of $\delta y / \delta x$ as δx tends to zero is $3x^2$, i.e. $dy/dx = 3x^2$.

We may now calculate the value of the gradient for any value of x that we care to choose, for example if $x = 2$ then the slope of the curve at this point is $3(2)^2 = 12$; if $x = 1$, the slope is 3; if $x = 0$, the slope is 0, etc.

There is a notation for specifying the differential coefficient for a particular value of x. If $f'(x)$ represents the differential coefficient with respect to x, then $f'(2)$, $f'(a)$ and $f'(0)$ denote the differential coefficient with respect to x for x having the values 2, a and 0.

So far we have considered functions of an independent variable x and used the notation $y = f(x)$. It is of course possible for us to use any symbols we like instead of x and y. Thus if time is the independent variable we may write

$$y = f(t)$$

and use the notation

$$\frac{dy}{dt}$$

to denote the differential coefficient of y with respect to t. Likewise we may write

$$\Psi = f(\theta)$$

and use the notations

$$\frac{d\Psi}{d\theta}, \; D\Psi, \; \frac{d}{d\theta} f(\theta) \text{ etc.}$$

to denote the differential coefficient of Ψ with respect to θ.

Example I

If $z = f(t) = 6t^2 + t + 3$ find $f'(t)$, $f'(3)$ and $f'(a)$.

By definition

$$f'(t) = \underset{\delta t \to 0}{\text{Lt}} \left\{ \frac{\delta z}{\delta t} \right\} = \underset{\delta t \to 0}{\text{Lt}} \left\{ \frac{f(t + \delta t) - f(t)}{\delta t} \right\}$$

For the case of this example we may therefore write:

$$f'(t) = \underset{\delta t \to 0}{\text{Lt}} \left\{ \frac{6(t + \delta t)^2 + (t + \delta t) + 3 - 6t^2 - t - 3}{\delta t} \right\}$$

$$= \underset{\delta t \to 0}{\text{Lt}} \left\{ \frac{6(2t + \delta t)(\delta t) + \delta t}{\delta t} \right\}$$

As δt approaches zero the expression in the brackets approaches the value $12t + 1$. Therefore we may write

$$f'(t) = 12t + 1.$$

From this expression we obtain

$$f'(3) = 12(3) + 1 = 36 + 1 = 37,$$

and

$$f'(a) = 12a + 1.$$

Example II

If $y = \frac{1}{x}$ find dy/dx.

$$\frac{dy}{dx} = \underset{\delta x \to 0}{\text{Lt}} \frac{\left\{ \frac{1}{x + \delta x} - \frac{1}{x} \right\}}{\delta x}$$

$$= \underset{\delta x \to 0}{\text{Lt}} \left\{ \frac{x - (x + \delta x)}{\delta x (x)(x + \delta x)} \right\}$$

As δx gets smaller and smaller, the limit tends to the value $-1/x^2$, so that we may write

$$\frac{dy}{dx} = -\frac{1}{x^2}.$$

Exercises 5.5

(1) If $f(x) = 3x^2$
find $f'(x)$, $f'(3)$ and $f'(0)$.

(2) If $\tau = \theta^3$

find $\dfrac{d\tau}{d\theta}$.

(3) If $y = 4x + 3$

find $\dfrac{dy}{dx}$.

(4) If $y = \dfrac{x^2}{5}$

find y'.

(5) If $z = \dfrac{1}{t + 2}$
find $z'(t)$ and $z'(-4)$.

5.6   THE DIFFERENTIAL COEFFICIENT OF THE SUM OR DIFFERENCE OF TWO FUNCTIONS

It must be obvious, if you have carried out the exercises of the previous section, that some of the problems of differentiation could become quite tedious if each problem had to be solved by the methods that we have discussed so far. However, there are many tricks and dodges that can be employed to make differentiation so much easier. The first method to be discussed is the case where we have the sum or difference of two functions to differentiate. We go through the proof here in its most general form for in this way we do not limit ourselves to any particular pair of functions.

Suppose

$$y = f(x) = u + v$$

where $u = f_1(x)$ and $v = f_2(x)$. That is both u and v are functions of x. We now want to know how to find dy/dx in terms of u and v or their derivatives.

First of all we consider x to be increased by a small amount to become $x + \delta x$. In this case u increases to $u + \delta u$ so that

$$u + \delta u = f_1(x + \delta x)$$

and also v becomes $v + \delta v$ so that

$$v + \delta v = f_2(x + \delta x).$$

By definition

$$\frac{dy}{dx} = \underset{\delta x \to 0}{Lt} \left\{ \frac{(y + \delta y) - y}{\delta x} \right\}$$

$$= \underset{\delta x \to 0}{Lt} \left\{ \frac{f_1(x + \delta x) + f_2(x + \delta x) - f_1(x) - f_2(x)}{\delta x} \right\}$$

$$= \underset{\delta x \to 0}{Lt} \left\{ \frac{(u + \delta u) + (v + \delta v) - u - v}{\delta x} \right\}$$

$$= \underset{\delta x \to 0}{Lt} \left\{ \frac{\delta u + \delta v}{\delta x} \right\} = \underset{\delta x \to 0}{Lt} \frac{\delta u}{\delta x} + \underset{\delta x \to 0}{Lt} \frac{\delta v}{\delta x}$$

$$= \frac{du}{dx} + \frac{dv}{dx}.$$

Thus we can summarise this result as follows:

If

$$y = u + v$$

then

$$\frac{dy}{dx} = \frac{du}{dx} + \frac{dv}{dx}$$

provided that both u and v are functions of the independent variable x.

We may extend this result to cover the cases where y is the sum of several functions of x. By reasoning similar to the case above, we may show that if

$$y = f_1(x) + f_2(x) + f_3(x) + + \ldots$$

then

$$y' = f_1'(x) + f_2'(x) + f_3'(x) + + \ldots \tag{5.7}$$

Also we may use similar reasoning to show that if

$$y = f_1(x) - f_2(x)$$

then

$$y' = f_1'(x) - f_2'(x). \tag{5.8}$$

Example

To differentiate with respect to x

$$\frac{1}{x+1} + \frac{1}{x-1}$$

We put

$$u = \frac{1}{x+1}$$

and

$$v = \frac{1}{x-1}.$$

We now differentiate u and v separately.

$$\frac{du}{dx} = \underset{\delta x \to 0}{\mathrm{Lt}} \left\{ \frac{\dfrac{1}{x + \delta x + 1} - \dfrac{1}{x + 1}}{\delta x} \right\}$$

$$= \underset{\delta x \to 0}{\mathrm{Lt}} \left\{ \frac{(x + 1) - (x + \delta x + 1)}{\delta x (x + \delta x + 1)(x + 1)} \right\}$$

$$= \underset{\delta x \to 0}{\mathrm{Lt}} \frac{-\delta x}{\delta x (x + \delta x + 1)(x + 1)}$$

$$= \frac{-1}{(x + 1)^2}.$$

Likewise, we may show that, if

$$v = \frac{1}{x - 1},$$

then

$$v' = \frac{-1}{(x - 1)^2}.$$

We are now in a position to write

$$y' = u' + v'$$

$$= -\left\{ \frac{1}{(x + 1)^2} + \frac{1}{(x - 1)^2} \right\}$$

$$= -\left\{ \frac{(x - 1)^2 + (x + 1)^2}{(x + 1)^2 (x - 1)^2} \right\}$$

$$= -2\left\{ \frac{x^2 + 1}{(x^2 - 1)^2}. \right\}$$

Exercises 5.6

Use the answers to the problems and exercises of Section 5.5 to find the differential coefficients of the following functions:

(1) $x^3 + 3x^2 + 4x + 3$

(2) $\dfrac{1}{x + 2} - \dfrac{1}{x}$

(3) $\dfrac{x^2}{5} + 4x + 3.$

5.7   THE DIFFERENTIAL COEFFICIENT OF THE PRODUCT OF TWO FUNCTIONS

If we have two functions u and v which are both functions of x then their product uv is also a function of x.

i.e.

$$y = f(x) = uv.$$

If we now change the value of x to $x + \delta x$ then u becomes $u + \delta u$, v becomes $v + \delta v$ and y becomes $y + \delta y$.

Hence

$$y + \delta y = (u + \delta u)\,(v + \delta v),$$

and

$$\delta y = (u + \delta u)\,(v + \delta v) - uv$$
$$= u\delta v + v\delta u + \delta u\delta v.$$

We may now write

$$\frac{dy}{dx} = \operatorname*{Lt}_{\delta x \to 0} \left\{ \frac{u\,\delta v}{\delta x} + \frac{v\,\delta u}{\delta x} + \frac{\delta u\,\delta v}{\delta x} \right\}$$

$$= \operatorname*{Lt}_{\delta x \to 0} \left\{ (u + \delta u)\frac{\delta v}{\delta x} + v\,\frac{\delta u}{\delta x} \right\}$$

As δx tends to zero then δu will become very small compared with u, and so we may write

$$\frac{dy}{dx} = \operatorname*{Lt}_{\delta x \to 0} \left\{ u\,\frac{\delta v}{\delta x} \right\} + \operatorname*{Lt}_{\delta x \to 0} \left\{ v\,\frac{\delta u}{\delta x} \right\}$$

$$= u\,\frac{dv}{dx} + v\,\frac{du}{dx}.$$

$$(5.9)$$

In alternative notation, we may write

$$y' = uv' + u'v.$$

We may put this result into words as follows:

'The differential coefficient of the product of two functions is equal to the second function multiplied by the differential coefficient of the first function plus the first function multiplied by the differential coefficient of the second'.

Example I

To find the differential coefficient of

$$y = (x + 2)(x^2 + 3).$$

Put $u = x + 2$ and $v = x^2 + 3$.

$$u' = \underset{\delta x \to 0}{\text{Lt}} \left\{ \frac{(x + \delta x + 2) - (x + 2)}{\delta x} \right\}$$

$$= 1$$

$$u' = \underset{\delta x \to 0}{\text{Lt}} \left\{ \frac{(x + \delta x)^2 + 3 - x^2 - 3}{\delta x} \right\}$$

$$= \underset{\delta x \to 0}{\text{Lt}} \left\{ \frac{2x \, \delta x + (\delta x)^2}{\delta x} \right\}$$

$$= 2x.$$

According to expression (5.9) derived above we may write

$$y' = uv' + u'v$$
$$= (x + 2)(2x) + (x^2 + 3)(1)$$
$$= 2x^2 + 4x + x^2 + 3$$
$$= 3x^2 + 4x + 3.$$

Example II

To find y' if

$$y = \frac{x^2 + x}{x - 1}.$$

Put $u = x^2 + x$

$$v = \frac{1}{x - 1}.$$

Then

$$u' = \underset{\delta x \to 0}{\text{Lt}} \left\{ \frac{(x + \delta x)^2 + (x + \delta x) - x^2 - x}{\delta x} \right\}$$

$$= \underset{\delta x \to 0}{\text{Lt}} \left\{ \frac{2x \, \delta x + (\delta x)^2 + \delta x}{\delta x} \right\}$$

$$= 2x + 1.$$

The result obtained in the last section can be used so that we may write

$$v' = \frac{-1}{(x-1)^2}.$$

Using the result

$$y' = uv' + u'v,$$

we write

$$y' = (x^2 + x)\frac{-1}{(x-1)^2} + \frac{(2x+1)}{(x-1)}$$

$$= \frac{-x^2 - x + 2x^2 - x - 1}{(x-1)^2}$$

$$= \frac{x^2 - 2x - 1}{(x-1)^2}.$$

Consider now the case where $u = v$. Instead of putting $y = uv$, we write $y = u^2$. Likewise when we differentiate, instead of putting

$$y' = uv' + u'v$$

we write

$$y' = u.u' + u'.u \qquad\qquad (5.10)$$

$$= 2u.u'.$$

Example III

To differentiate

$$y = \left\{\frac{x^2 + x}{x - 1}\right\}^2$$

We put

$$u = \left(\frac{x^2 + x}{x + 1}\right),$$

and, from the result of the previous example, we write

$$u' = \frac{x^2 - 2x - 1}{(x - 1)^2} .$$

By the result of expression (5.10) above we write

$$y' = \frac{2(x^2 + x)(x^2 - 2x - 1)}{(x - 1)^3} .$$

Suppose that we now have the function

$$y = uvw$$

where u, v and w are all functions of x. In order to differentiate this we write

$$y = u(vw)$$

and we treat the product vw as a single function of x. We use expression (5.9) to write

$$y' = u(vw)' + u'(vw).$$

However, $(vw)'$ is the differential coefficient of the product of two functions and we write, again using expression (5.9),

$$(vw)' = vw' + v'w,$$

so that

$$y' = uvw' + uv'w + u'vw. \qquad (5.11)$$

In the special case we might have

$$u = v = w$$

so that

$$y = u^3$$

and

$$y' = 3u^2u'. \qquad (5.12)$$

Finally, for the case where

$$y = uvw\theta\lambda\psi v,$$

which is the product of several functions of x, we may write,

$$y' = u'vw\theta\lambda\psi v + uv'w\theta\lambda\psi v + uvw'\theta\lambda\psi v + \\ uvw\theta'\lambda\psi v + uvw\theta\lambda'\psi v + uvw\theta\lambda\psi'v + \\ uvw\theta\lambda\psi v'.$$

(5.13)

Exercises 5.7

Differentiate the following functions with respect to x:

(1) $(x^2 + 1)(x)$

(2) $(x)(1 - x)$

(3) $(x + 2)(x + 1)$

(4) $\dfrac{x + 2}{x + 1}$

(5) $(x)(x + 1)(x + 2)$

(6) $(x + 4)^3$

(7) $\dfrac{(x + 1)^2}{(x - 1)}$

(8) $(x + 3)^2(x - 1)^3$

(9) $\dfrac{1}{(x)(x - 1)}$

(10) $x^3 - 3x^2 + 3x - 1$

(11) $x^4 - 2x^2 - 1$

(12) $x^3 - 6x^2 - 9x$

(13) $x^2 + (a - b)x - ab$

(14) $(x + a)^2(x - b)$.

5.8 DIFFERENTIATION INVOLVING A CONSTANT

Consider the equation

$$y = b$$

where b is a constant. This will be recognised as the equation of a straight line parallel to the X-axis and it is depicted in Fig. 3.15 (ii). It is seen from this figure that any change in the value of x to $x + \delta x$ produces no change in the value of y. Thus δy is always zero and thus the slope of the line is zero. In this case then

$$y' = 0.$$

(5.14)

We now use this result to consider the differentiation of the function

$$y = b \cdot u$$

where b is a constant and u is any function of x. Let us put

$$v = b$$

and use the result of the last section to write

$$y' = vu' + v'u. \tag{5.15}$$

In this case $v = b = $ constant and therefore $v' = 0$. Substitution of these values into equation (5.15) gives the result

$$y' = b \cdot u'. \tag{5.16}$$

In words we may say 'The differential coefficient of a constant times a function is equal to the constant times the differential coefficient of the function'.

5.9   TO SHOW THAT $\frac{d(x^n)}{dx} = nx^{n-1}$

(a) n is a positive integer

Let us first of all take the case for $n = 1$.
If

$$y = x,$$

then by definition

$$\frac{dy}{dx} = \operatorname*{Lt}_{\delta x \to 0} \frac{(x + \delta x) - x}{\delta x}$$

$$= 1.$$

In section 5.7 we showed that if

$$y = u^2,$$

then

$$y' = 2uu'.$$

However, if

$$y = x^2,$$

then

$$u = x,$$

and

$$u' = 1,$$

so that

$$y' = 2x.$$

Likewise for $n = 3$ we can use the result of equation (5.12) to show that

$$\frac{d(x^3)}{dx} = 3x^2.$$

To summarise the above results we have shown that

(i) if $y = x$, $y' = 1$,
(ii) if $y = x^2$, $y' = 2x$,
(iii) if $y = x^3$, $y' = 3x^2$.

Thus for values of $n = 1$, 2, or 3 we may write

$$\text{if } y = x^n,$$

$$y' = nx^{n-1}.$$

We now need to prove this result to be true for *any* positive value of n. In order to do this we use what is known as a *method of induction*.

We know that

$$\text{if } y = x^n$$

then

$$y' = nx^{n-1}$$

holds for $n = 1$, 2, and 3. For any one of these values of n let us consider the case for $n + 1$.

Thus

$$y = x^{n+1}$$

$$= (x^n)(x).$$

By the formula for differentiation of a product we write

$$y' = x^n \frac{d(x)}{dx} + x \frac{d(x^n)}{dx} \tag{5.17}$$

We know that

$$\frac{d(x)}{dx} = 1,$$

and that, for $n = 1$, 2 or 3,

$$\frac{d(x^n)}{dx} = nx^{n-1}.$$

So we may put equation (5.17) in the form

$$y' = x^n(1) + (x)(nx^{n-1})$$
$$= (n+1)x^n$$
$$= \frac{d(x^{n+1})}{dx}.$$

Thus it is seen that if the result holds for a particular value of n, it also holds for $n + 1$. By continuing the reasoning we can show that the result is true for $n + 2$, $n + 3, n + 4$ or to any integral value that we care to choose. We know that the result holds for $n = 1$, 2 or 3, so it also holds for $n = 4, 5, 6 \ldots$, in fact for any positive integer.

Example I

To differentiate $y = x^{10}$.
According to the formula

$$\frac{d(x^n)}{dx} = nx^{n-1}.$$

In this case, $n = 10$, so we write

$$\frac{d(x^{10})}{dx} = 10x^9.$$

(b) n is a negative integer.

Given that n is a negative integer, we may write

$$x^n = x^{-m}$$

where $m = -n$ and m is a positive integer.

If we write the relationship

$$y = x^{-m} x^m = 1,$$ (5.18)

we can write immediately

$$y' = 0.$$ (5.19)

But

$$y' = \frac{d(x^{-m} x^m)}{dx} = (x^m) \frac{d(x^{-m})}{dx} + (x^{-m}) \frac{d(x^m)}{dx}.$$

However, from equations (5.18) and (5.19) this relationship is zero and we write

$$(x^m) \frac{d(x^{-m})}{dx} + (x^{-m}) \frac{d(x^m)}{dx} = 0.$$ (5.20)

Now m is a positive integer so that we may write

$$\frac{d(x^m)}{dx} = mx^{m-1},$$

and expression (5.20) takes the form

$$(x^m) \frac{d(x^{-m})}{dx} + (x^{-m})(mx^{m-1}) = 0,$$

which is rearranged to give

$$\frac{d(x^{-m})}{dx} = -mx^{-m-1}.$$

But $m = -n$ so we may write

$$\frac{d(x^n)}{dx} = nx^{n-1}.$$ (5.21)

Thus we have proved the relationship for any negative integer.

Example II

To differentiate $y = x^{-2}$.

In this case $n = -2$, so that we may use the formula (5.21) to write

$$y' = -2x^{-3}.$$

(c) $n = p/q$ where p and q are integers

In section 5.7 we showed that if

$$y = u^2$$

then

$$y' = 2uu',$$

and if

$$y = u^3$$

then

$$y' = 3u^2u'.$$

Now if

$$y = u^4$$

we may write

$$y = u^3 u$$

and

$$\begin{aligned} y' &= u(u^3)' + u'u^3 \\ &= u(3u^2u') + u'u^3 \\ &= 4u^3u'. \end{aligned} \tag{5.22}$$

We now wish to show that, for any positive integer n and any function $u(x)$, if

$$y = u^n,$$

then

$$\frac{dy}{dx} = n(u^{n-1})u'. \tag{5.23}$$

In order to do this we use a method of induction very similar to that which we have already used. We know that expression (5.23) is true for values of $n = 1, 2,$

3 and 4. We now put

$$y = u^{n+1}$$

where $n = 1, 2, 3,$ or 4. In order to differentiate we put it in the form

$$y = (u)(u^n),$$

and differentiate the product thus:

$$y' = (u)(u^n)' + (u')(u^n).$$

We use expression (5.23) to write

$$y' = (u)[n(u^{n-1})u'] + u'(u^n)$$
$$= (n + 1)u^n u'$$

(5.24)

Thus it is seen that if the result holds for a particular value of n, it also holds for $n + 1$. By continuing the reasoning we can show that the result is true for $n + 2$, $n + 3, n + 4$ etc. In other words, the result holds for any positive integral value of n.

We now use this result, which we have considered in some detail, to differentiate the function

$$y = x^{p/q}$$

(5.25)

where p and q are integers. We may raise the whole expression to the power of q and obtain

$$y^q = (x^{p/q})^q = x^p.$$

Since y is a function of x we may use the result of equation (5.23) to obtain

$$\frac{d(y^q)}{dx} = qy^{q-1}\frac{dy}{dx} = q(x^{p/q})^{q-1}\frac{dy}{dx},$$

but

$$y^q = x^p,$$

therefore

$$\frac{d(y^q)}{dx} = px^{p-1}$$

and

$$q(x^{p/q})^{q-1} \frac{dy}{dx} = px^{p-1}.$$

Rearrangement of this expression yields

$$\frac{dy}{dx} = \frac{p}{q} x^{(p/q-1)},$$

and if we put $p/q = n$ we obtain

$$\frac{dy}{dx} = nx^{n-1}$$

and so far we can say that this holds good where n is any integer or rational fraction.

Example III

To differentiate $y = x^{1/2}$.

In this case $n = \frac{1}{2}$ and $(n - 1) = -\frac{1}{2}$, so

$$y' = \frac{1}{2} x^{-1/2}.$$

Example IV

To differentiate

$$y = x^{-2/3}.$$

In this case $n = -\frac{2}{3}$ and $(n - 1) = -\frac{5}{3}$, so

$$y' = -(\frac{2}{3}) x^{-5/3}.$$

(d) n is an irrational number

Suppose n is a recurring decimal or some such number as π or e. In order to show by means of a rigorous proof that

$$\frac{d(x^n)}{dx} = nx^{n-1}$$

the more advanced textbooks must be consulted. However, we can say at this stage that whatever the irrational number, it can usually be fairly well approximated to the form p/q where p and q are integers and then resort to the proof given above. If this is accepted by the reader, we may state that

$$\frac{d(x^n)}{dx} = nx^{n-1}$$

for n having any real value.

Example V

To differentiate $y = x^{\pi/2}$.

In this case $n = \pi/2$ and $(n - 1) = \pi/2 - 1$, so

$$y' = \frac{\pi}{2} x^{(\pi/2-1)}.$$

Example VI

To differentiate $y = (x^3 + 3)^{-1/2}$.
If we put

$$u = (x^3 + 3)$$

and

$$n = -\tfrac{1}{2},$$

then

$$y = u^n.$$

By expression (5.23) we may write

$$y' = n(u^{n-1})u'.$$

But

$$u' = \frac{d}{dx}(x^3 + 3)$$

$$= \frac{d}{dx}(x^3) + \frac{d(3)}{dx}$$

$$= 3x^2 + 0,$$

$$= 3x^2.$$

Also

$$(n - 1) = -\tfrac{3}{2}.$$

We are now able to write out the answer as follows:

$$y' = -\tfrac{1}{2}(x^3 + 3)^{-3/2}(3x^2)$$

$$= -(\tfrac{3}{2})x^2(x^3 + 3)^{-3/2}.$$

Exercises 5.9

Differentiate the following functions with respect to x:

(1) $x^4 + 3x^2$

(2) $(x + 1)^3$

(3) $x^2(x^2 + 2)^3$

(4) $\dfrac{1}{2x}$

(5) $x^{-1/3}$

(6) x^π

(7) $(x + 1)^{1/2}$

(8) $(x^2 + 1)^{-e}$

(9) $\dfrac{x}{(x^2 - 1)^{1/2}}$

(10) $(x - 1)^{-4}$

(11) $x^3(x-1)^2$ (12) x^{-5}

(13) $(3x-2)^{-3}$ (14) $x^{3/2}$

(15) $(x+2)^{\pi/5}$ (16) $(x^{1/2}-3)x^{-1/3}$

(17) $\dfrac{x^{1/2}}{(4x-1)^{1/2}}$ (18) $(x+3)^{3/2}$

(19) $(x+1)^2(x-1)^e$ (20) $(a+x)^{-2}$
                                          [a is a constant]

(21) $x^{1/2}+x^{-1/2}$ (22) $(x^3+2x+1)^{-1/2}$

(23) $\dfrac{x^{1/2}}{x^{1/2}-1}$ (24) $\dfrac{(a+x)^{1/2}}{(b-x)^{1/2}}$

                                          [a and b are
                                          constants]

(25) If, at constant temperature, the pressure (p) and volume (v) of a gas are connected by the relation

$$p \cdot v = \text{constant},$$

show that the cubical elasticity $(-v\mathrm{d}p/\mathrm{d}v)$ is equal to p.

5.10 DIFFERENTIATION OF e^x

By definition

$$e^x = 1 + x + \frac{x^2}{2!} + \frac{x^3}{3!} + \frac{x^4}{4!} + + + \tag{5.26}$$

We differentiate each term separately by using the formula obtained in the previous chapter

$$\frac{\mathrm{d}(x^n)}{\mathrm{d}x} = nx^{n-1}.$$

$$\frac{d(e^x)}{dx} = 0 + 1 + \frac{2x}{2!} + \frac{3x^2}{3!} + \frac{4x^3}{4!} + + + \tag{5.27}$$

In Section 2.1 we saw that

$$n! = n(n - 1)!,$$

but this equation can be put into the form

$$\frac{n}{n!} = \frac{1}{(n - 1)!}.$$

We can use this formula to simplify equation (5.27) thus:

$$\frac{d(e^x)}{dx} = 1 + x + \frac{x^2}{2!} + \frac{x^3}{3!} + + +$$

$$= e^x.$$

This is a very important and interesting result; the differential coefficient of e^x is equal to e^x.

We now go on to differentiate two common forms of the exponential series:

(a) To differentiate $y = e^{-x}$.

We expand the function

$$e^{-x} = 1 - x + \frac{x^2}{2!} - \frac{x^3}{3!} + \frac{x^4}{4!} - \cdots \tag{5.28}$$

We differentiate term by term to obtain

$$\frac{d(e^{-x})}{dx} = 0 - 1 + \frac{2x}{2!} - \frac{3x^2}{3!} + \frac{4x^3}{4!} - \cdots$$

This is further simplified to obtain

$$\frac{d(e^{-x})}{dx} = -\left\{1 - x + \frac{x^2}{2!} - \frac{x^3}{3!} + \cdots\right\}$$

$$= -e^{-x} \tag{5.29}$$

(b) To differentiate $y = e^{ax}$ where a is a constant.

As before we expand the function and differentiate term by term:

$$e^{ax} = 1 + ax + \frac{(ax)^2}{2!} + \frac{(ax)^3}{3!} + + +$$

$$\frac{d(e^{ax})}{dx} = 0 + a + a^2 x + \frac{a^3 x^2}{2!} + + +$$

$$= a\left(1 + ax + \frac{(ax)^2}{2!} + +\right)$$ (5.30)

$$= a\, e^{ax}.$$

Example I

To differentiate $y = e^{\frac{1}{2}x}.$

In this case $a = \frac{1}{2}$, so that

$$y' = \tfrac{1}{2}e^{\frac{1}{2}x}.$$

Example II

To differentiate $y = 6 \exp[-4t]$.

In this case $a = -4$, so that
$$y' = 6(-4 \exp[-4t])$$
$$= -24 \exp[-4t].$$

Exercises 5.10

(1) Differentiate the following functions with respect to x:

   (a) $(x + 4)e^x$

   (b) $5e^x + e^{-x}$

   (c) $e^x\left[x + \dfrac{1}{x}\right]$

   (d) $x^2 \exp[2x]$

   (e) $\exp[a + bx]$ (a and b are constants).

(2) *A* drug is infused into the circulation of an animal at time $t = 0$. The concentration of the drug at any subsequent time t is given by the relationship

$$C(t) = G_1 \exp[-k_1 t] + G_2 \exp[-k_2 t]$$

where $C(t)$ is the concentration at a time t; G_1, k_1, G_2 and k_2 are constants. If V is the volume of the circulation, determine the rate of change of the amount of the substance in the circulation. Express your answer in a formula which contains the given constants.

(3) A bacterial population has a size $A \exp[t/\tau]$ where A and τ are constants. What is the rate of growth at time $t = \tau$? Express your answer in terms of the constants.

5.11   TO DIFFERENTIATE A FUNCTION OF A FUNCTION

Suppose we have a function of x

$$u = f(x) \tag{5.31}$$

and also suppose that the function u is incorporated in another function such that

$$y = g(u) \tag{5.32}$$

then from these two expressions we may write

$$y = g[f(x)]. \tag{5.33}$$

For example we might have

$$u = (x^2 + 1)$$

and

$$y = \exp(u),$$

then

$$y = \exp[x^2 + 1].$$

This function is known as a *function of a function* because $\exp[u]$ is a function of u and u is a function of x.

We now wish to differentiate expression (5.33) with respect to x. We consider expressions (5.31) and (5.32). If x is changed by a small amount to $x + \delta x$, then u will change to $u + \delta u$ and thus y will change to $y + \delta y$. We may also write

$$\frac{\delta y}{\delta x} = \frac{\delta y}{\delta u} \cdot \frac{\delta u}{\delta x}.$$

As δx gets smaller and smaller, then δu will get smaller and smaller, because u is a function of x, and so $\delta u/\delta x$ will tend to the limit du/dx. At the same time, since δu gets smaller and smaller, δy will get smaller and smaller also, because y is a function of u, and so $\delta y/\delta u$ will tend to the limit dy/du. Thus we may write

$$\frac{dy}{dx} = \underset{\delta x \to 0}{\text{Lt}} \frac{\delta y}{\delta x}$$

$$= \underset{\delta x \to 0}{\text{Lt}} \frac{\delta y}{\delta u} \cdot \frac{\delta u}{\delta x} \qquad (5.34)$$

$$= \frac{dy}{du} \cdot \frac{du}{dx}.$$

Example I

To differentiate $y = \exp[x^2 + 1]$.

Put $u = (x^2 + 1)$
and $y = \exp[u]$.

Differentiation of the two functions gives

$$\frac{du}{dx} = 2x,$$

and

$$\frac{dy}{du} = \exp[u].$$

From expression (5.34) we may write

$$\frac{dy}{dx} = \frac{dy}{du} \cdot \frac{du}{dx}$$

$$= (\exp[u])(2x)$$

$$= 2x \cdot \exp[x^2 + 1].$$

Example II

To differentiate
$$y = \frac{1}{(x^2 + x + 1)^2}.$$

In this example we put

$$u = (x^2 + x + 1)$$

and

$$y = u^{-2}.$$

We differentiate these two functions

$$\frac{du}{dx} = (2x + 1)$$

and

$$\frac{dy}{du} = -2u^{-3}.$$

We use expression (5.34) to give

$$\frac{dy}{dx} = (-2u^{-3})(2x + 1)$$

$$= \frac{-2(2x + 1)}{(x^2 + x + 1)^3}.$$

Exercises 5.11

Differentiate the following functions with respect to x:

(1) $(2x + 4)^3$ (2) $(2x + 1)^{-2}$

(3) $(2x + 1)^2 + (2x + 1)$ (4) $(1 + x)(1 - x)^3$

(5) $\dfrac{(1 + x)^3}{1 - x}$ (6) $(x + 3)^2 + x$

(7) $(x^2 + 1)^2 + \dfrac{1}{(x^2 + 1)}$ (8) $\exp[-x^4]$

(9) $\exp[(x^3 + 2x)^{1/2}]$

5.12   TO DIFFERENTIATE THE QUOTIENT OF TWO FUNCTIONS

Suppose we have

$$y = \frac{u}{v}$$

where u and v are functions of x.

The expression may be put in the following form

$$y = u \,.\, v^{-1},$$

and, for the purposes of differentiation, we treat it as the product of two functions:

$$\frac{dy}{dx} = u \,.\, \frac{d(v^{-1})}{dx} + (v^{-1}) \frac{du}{dx}. \tag{5.35}$$

We now use the result of the previous section and write

$$\frac{d(v^{-1})}{dx} = \frac{d(v^{-1})}{dv} \,.\, \frac{dv}{dx}.$$

But

$$\frac{d(v^{-1})}{dv} = -\frac{1}{v^2},$$

and thus

$$\frac{d(v^{-1})}{dx} = -\frac{1}{v^2}\frac{dv}{dx}.$$

We substitute this result into expression (5.35) to obtain:

$$\frac{dy}{dx} = -\frac{u}{v^2}\frac{dv}{dx} + \frac{1}{v}\frac{du}{dx}$$

$$= \frac{v\dfrac{du}{dx} - u\dfrac{dv}{dx}}{v^2} \tag{5.36}$$

Example I

To differentiate

$$y = \frac{(x^2 - 2x + 4)}{(x^2 + 2x + 4)}.$$

We put

$$u = (x^2 - 2x + 4)$$

and

$$v = (x^2 + 2x + 4)$$

These two functions are easily differentiated to give

$$u' = (2x - 2)$$

and

$$v' = (2x + 2).$$

We substitute these expressions into equation (5.36) to obtain

$$y' = \frac{(x^2 + 2x + 4)(2x - 2) - (x^2 - 2x + 4)(2x + 2)}{(x^2 + 2x + 4)^2}$$

$$= \frac{4x^2 - 16}{(x^2 + 2x + 4)^2}$$

$$= \frac{4(x - 2)(x + 2)}{(x + 2)^4}$$

$$= \frac{4(x - 2)}{(x + 2)^3}.$$

Exercises 5.12

Differentiate the following functions with respect to x:

(1) $\dfrac{x + 1}{2x - 1}$

(2) $\dfrac{x^2 + 3x}{x + 2}$

(3) $\dfrac{x^3}{x^2 - 1}$

(4) $\dfrac{3x}{x^2 + 4}$

(5) $\dfrac{x^2}{3 - x}$

(6) $\dfrac{7x}{3 - x^2}$

(7) $(x^2 - 2)^{-1}$ (8) $\dfrac{(x + 1)^2}{(x - 1)}$

(9) $\dfrac{3x + 4}{(x - 1)^2}$ (10) $\dfrac{(3x^2 + 1)^4}{(x^2 - 1)^{10}}$

(11) $\dfrac{8x^2}{4x + 1}$

5.13   DIFFERENTIATION OF INVERSE FUNCTIONS

If y is a function of x, then generally x is a function of y.

For example, if we write

$$y = x^{1/2},$$

then we may also write

$$x = y^2.$$

In general if

$$y = f(x),$$

then we may write

$$x = \phi(y)$$

and $\phi(y)$ is known as the *inverse function*.

Consider now the result obtained for the differentiation of a function of a function

$$\frac{dy}{dx} = \frac{dy}{du} \cdot \frac{du}{dx}.$$

Now suppose we take the case where $y = x$. We may now write

$$\frac{dx}{dx} = \frac{dx}{du} \cdot \frac{du}{dx}.$$

But

$$\frac{dx}{dx} = 1$$

and therefore

$$\frac{dx}{du}\cdot\frac{du}{dx} = 1$$

and

$$\frac{dx}{du} = \frac{1}{du/dx}.$$

(5.37)

Example I

To find dx/dy for $y = x^{1/5}$.

We can do this in two ways

(1) put the equation in the form

$$x = y^5$$

and write

$$\frac{dx}{dy} = 5y^4 = 5x^{4/5};$$

(2) write

$$\frac{dy}{dx} = \tfrac{1}{5} x^{-4/5}$$

and invert the function, according to expression (5.37), and obtain

$$\frac{dx}{dy} = 5x^{4/5}.$$

Example II

Find dy/dx if

$$y^2 + 3y + x = 0.$$

First of all we write

$$x = -y^2 - 3y,$$

and differentiate with respect to y:

$$\frac{dx}{dy} = -2y - 3.$$

We invert this relationship to obtain

$$\frac{dy}{dx} = -\frac{1}{2y + 3}.$$

Exercises 5.13

Find dy/dx for the following functions

(1) $y^3 + 3y = x + 1$

(2) $x(y + a) = y^2$ [a = const.]

(3) $(2y + 1)(y + 4) = x$

(4) $xy + x + y + y^3 = 0.$

5.14   DIFFERENTIATION OF LOGARITHMS

If

$$y = \ln x,$$

then we know from section 1.3 that

$$x = e^y.$$

If we differentiate with respect to y we obtain

$$\frac{dx}{dy} = e^y$$

and therefore

$$\frac{dy}{dx} = \frac{1}{e^y} = \frac{1}{x}. \tag{5.38}$$

Thus we have obtained the result that the differential coefficient of $\ln x$ is $1/x$. This result is now used in the differentiation of more complicated functions involving logarithms. These are illustrated by way of the following examples:

Example I

To differentiate $y = \ln (x^2 + x^{1/2} + 2x)$.

We put

$$u = (x^2 + x^{1/2} + 2x),$$

and

$$y = \ln u.$$

So that

$$\frac{du}{dx} = (2x + \tfrac{1}{2} x^{-1/2} + 2),$$

and

$$\frac{dy}{du} = \frac{1}{u}.$$

We use the rule for the differentiation of a function of a function to obtain

$$\frac{dy}{dx} = \frac{dy}{du} \cdot \frac{du}{dx}$$

$$= \frac{1}{u} (2x + \tfrac{1}{2} x^{-1/2} + 2)$$

$$= \frac{(2x + \tfrac{1}{2} x^{-1/2} + 2)}{(x^2 + x^{1/2} + 2x)}.$$

Example II

To differentiate $y = \log_a(x)$.

We first of all rearrange the equation to take the form

$$x = a^y \tag{5.39}$$

which we can do if we use the definition of a logarithm.

We now choose *a* constant *b* such that

$$a = e^b \text{ (i.e. } b = \ln a).$$

This is incorporated into expression (5.39) which now takes the form:

$$x = (e^b)^y = e^{by}.$$

Differentiation with respect to y gives

$$\frac{dx}{dy} = b\,e^{by}$$

This expression is now inverted to give

$$\frac{dy}{dx} = \frac{1}{b\,e^{by}}$$

but

$$e^{by} = x,$$

and

$$b = \ln a,$$

therefore

$$\frac{dy}{dx} = \frac{1}{x\ln(a)}.$$

Example III

To differentiate $y = a^x$.

We take natural logarithms of both sides

$$\ln(y) = x\ln(a),$$

which is rearranged to the form

$$x = \frac{\ln(y)}{\ln(a)}.$$

Differentiation with respect to y gives

$$\frac{dx}{dy} = \frac{1}{\ln(a)}\left(\frac{1}{y}\right).$$

This is inverted to give

$$\frac{dy}{dx} = y\ln(a)$$
$$= a^x\ln(a).$$

To differentiate a function which consists of a constant raised to the power of a variable, it is always a good rule to take natural logarithms in order to facilitate differentiation.

Exercises 5.14

Differentiate the following functions with respect to x:

(1) $a^{(x+2)}$

(2) $\ln (x^2 + 3)$

(3) $\log_{10}(x)$

(4) $\ln (x^3 + 2x)$

(5) $\ln (x^2 + 1)^3$

(6) $e^x + x \ln (x)$

(7) $x \ln (x) - x$

(8) $(x + 1)e^x + \ln (x^2)$

(9) $e^x \left\{ x + \dfrac{1}{x} \right\}$

(10) e^{3x}

(11) xe^x

(12) $(x + 1) \ln (x)$

(13) $\dfrac{\ln (x^2)}{x}$

(14) $\ln (3x)$

(15) $\ln \left\{ \dfrac{3x}{x^2 + 1} \right\}$.

5.15 DIFFERENTIAL COEFFICIENTS OF TRIGONOMETRIC FUNCTIONS

The trigonometric functions have not been dealt with to any great extent in this book. It is however possible to show that the following results can be obtained:

(1) If $y = \sin x$,

$$\frac{dy}{dx} = \cos x. \tag{5.40}$$

(2) If $y = \cos x$,

$$\frac{dy}{dx} = -\sin x. \qquad (5.41)$$

From these two results all the other trigonometric functions can be differentiated. In particular we note:

$$y = \tan x = \frac{\sin x}{\cos x}$$

$$\frac{dy}{dx} = \frac{\cos^2 x + \sin^2 x}{\cos^2 x} = \frac{1}{\cos^2 x} \qquad (5.42)$$

Note that $(\sin x)^2$ and $(\cos x)^2$ are usually written $\sin^2 x$ and $\cos^2 x$ respectively.

Example I

Differentiate $y = \sin(ax + b)$

where a and b are constants.

This is an example of a function of a function so we put

$$y = \sin u$$

where

$$u = (ax + b).$$

We differentiate as follows:

$$\frac{dy}{du} = \cos u$$

$$\frac{du}{dx} = a.$$

So that

$$\frac{dy}{dx} = \frac{dy}{du} \cdot \frac{du}{dx} = [\cos u] \cdot a = a\cos(ax + b).$$

Example II

Differentiate $y = \cos^2 x$.

Put

$$y = u^2$$

where

$$u = \cos x.$$

We differentiate as follows:

$$\frac{dy}{du} = 2u,$$

$$\frac{du}{dx} = -\sin x.$$

So that

$$\frac{dy}{dx} = \frac{dy}{du} \cdot \frac{du}{dx}$$

$$= 2u(-\sin x)$$

$$= -2 \sin x \cdot \cos x.$$

Exercises 5.15

Differentiate the following functions with respect to x:

(1) $\cos(nx)$
   (n is a constant)

(2) $\sin(nx)$

(3) $x + 2 \sin x$

(4) $\sin(x^3)$

(5) $\sin^2 x$

(6) $\sin x - \cos x$

(7) $\exp[\sin x]$

(8) $\ln(\sin x)$

(9) $(\sin x)^{1/2}$

(10) $\cos^n x$

(11) $(a + b \sin x)^2$

(12) $\cos^2 x \cdot \sin x$

5.16   TABLE OF STANDARD FORMS

In this section we summarise the results obtained in previous sections.

$$y = u \pm v; \frac{dy}{dx} = \frac{du}{dx} \pm \frac{dv}{dx}.$$

$$y = uv; \frac{dy}{dx} = u\frac{dv}{dx} + v\frac{du}{dx}.$$

$$y = f(u), u = f(x); \frac{dy}{dx} = \frac{dy}{du}.\frac{du}{dx}.$$

$$y = \frac{u}{v}; \frac{dy}{dx} = \frac{v\frac{du}{dx} - u\frac{dv}{dx}}{v^2}.$$

$$y = x^n; \frac{dy}{dx} = nx^{n-1}.$$

$$y = e^x; \frac{dy}{dx} = e^x.$$

$$y = \ln(x) \frac{dy}{dx} = \frac{1}{x}.$$

$$y = \sin(x) \frac{dy}{dx} = \cos(x).$$

$$y = \cos(x) \frac{dy}{dx} = -\sin(x).$$

Exercises 5.16

Differentiate the following functions with respect to the variable contained in each:

(1) $\sin^2\theta + \tan^2\theta$ (2) $\ln(1 - 2x + 2x^2)^{1/2}$

(3) $\sin(2x) . \cos(x)$ (4) $\cos(\pi + 3\theta)$

(5) $1/(3 \exp[2x])$ (6) $\ln[\tan(\tfrac{1}{2}\theta)]$

(7) $4^{3z} + \ln(4z^3)$ (8) $\ln(\cos\theta)$

(9) $(y + 1)^3 \exp[1 - y]$ (10) $\dfrac{v}{\cos(3y)}$

(11) $(0.2)^y$ (12) $\dfrac{y}{(p + 1)(p + 2)}$

(13) $\exp[\ln(x)]$ (14) $\exp[e^x]$

(15) $\log_{10}(e^x)$ (16) $\exp[\cos(\theta)]$.

6

HIGHER DIFFERENTIATION, MAXIMA AND MINIMA

6.1 DEFINITIONS

If a continuous function increases up to a certain value and then begins to decrease, that value is known as a maximum value. If the function decreases to a certain value and then begins to increase, that value is known as a minimum value.

More precisely, a function f(x) will have a maximum value f(a) at x = a if all

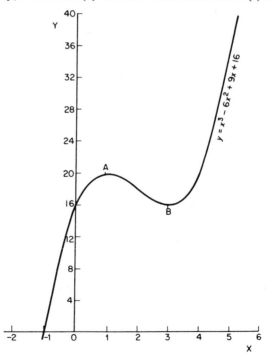

Fig. 6.1.

other values of f(x) in the interval $x = a - b$ to $x = a + c$ (b and c are positive quantities) are less than f(a). Similarly a function f(x) will have a minimum value f(a) at $x = a$ if all other values of f(x) in the interval $x = a - b$ to $x = a + c$ are greater than f(a).

A function may have any number of maxima and minima because the comparison is made with values of the function in the immediate neighbourhood of $x = a$. 'A maximum or minimum value will not necessarily be the greatest or least value that the function f(x) can attain. For instance in Fig. 6.1, which is a graph of $y = x^3 - 6x^2 + 9x + 16$, the ordinates at A and B are respectively maximum and minimum values but clearly these values do not represent the greatest and least possible values of the function.

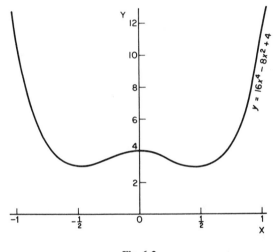

Fig. 6.2.

Some functions may have more than one maximum or minimum. For instance the function $y = 16x^4 - 8x^2 + 4$ (Fig. 6.2) has two minima and one maximum.

6.2 ALTERNATE MAXIMA AND MINIMA

Maxima and minima occur alternately in a function. This should be obvious because after any maximum the function decreases but before the next maximum is reached the function must be increasing. Therefore, between the two maxima there must be a point where the function ceases to decrease and starts to increase. This point will be a minimum for the function. Hence between any two consecutive maxima there is a minimum and between any two consecutive minima there is a maximum.

The curve in Fig. 6.2 illustrates this point. In order that the function may have a minimum value on either side of the Y-axis, the function has a maximum value at $x = 0$.

Exercises 6.2

Draw graphs of the following functions for the range of x indicated and determine the points where there are maxima and minima:

(1) $x^2 - 6x + 6$ (from $x = 0$ to $x = 6$)

(2) $x^3 - 12x + 1$ (from $x = -3$ to $x = 4$)

(3) $3x^3 + 15x^2 - 24x - 5$ (from $x = -5$ to $x = 1$)

(4) $x^3 - 3x^2 - 9x - 1$ (from $x = -4$ to $x = 4$).

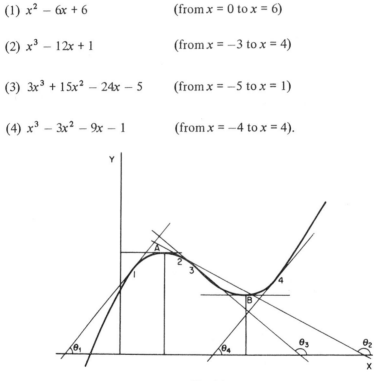

Fig. 6.3.

6.3   CONDITIONS FOR A MAXIMUM OR A MINIMUM

Consider the curve shown in Fig. 6.3. The ordinates at A and B represent maximum and minimum values of the function. The tangent at any point on the curve makes an angle θ with the X-axis. At the points A and B the tangents are parallel to the X-axis and therefore $\theta = 0$ and $\tan \theta = 0$. At other points on the curve near to the point A, the tangents are not parallel to the X-axis. Just before the point A the tangent makes an acute angle with the X-axis and tan θ has a positive value but decreases towards zero for points nearer to A. Points just after A on the curve have tangents which make obtuse angles with the X-axis and for these points tan θ has negative values. Hence in passing through a maximum tan θ changes from positive values to negative values. On the other hand points on the curve just before the minimum at point B have tangents which make obtuse angles with the X-axis and tan θ is negative. As the points get nearer to B the values of tan θ get progressively less negative until at B tan θ is zero. At points just after B the tangents make acute angles with the X-axis for which tan θ is positive. Hence in passing through a minimum value tan θ changes from negative values to positive values.

6.4 BEHAVIOUR OF THE DIFFERENTIAL COEFFICIENT AT MAXIMA AND MINIMA

In chapter 5 it was seen that the tangent to the curve $y = f(x)$ at a point (x, y) on the curve makes an angle θ with the X-axis and that $\tan 0 = dy/dx$.

If θ makes an acute angle with the X-axis then dy/dx has a positive value; if θ is zero then $dy/dx = 0$ and if θ is obtuse then dy/dx has a negative value. We may now restate the conclusions of the previous section in terms of the differential coefficient:

Just before the maximum value dy/dx has a positive value which decreases to

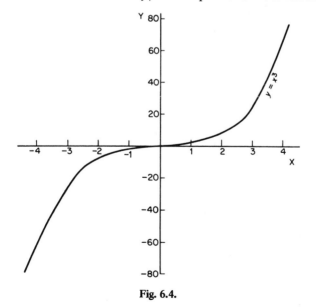

Fig. 6.4.

zero at the point of the maximum and then takes negative values. Hence in passing through a maximum dy/dx decreases and changes its sign from positive to negative. Likewise in passing through a minimum dy/dx increases and changes its sign from negative to positive.

The condition that $dy/dx = 0$ is not sufficient to establish a maximum or a minimum. It is possible that a function may increase up to a certain value where $dy/dx = 0$ and then continue to increase so that dy/dx is still positive. In other words dy/dx does not change its sign although it is zero at one point. For example the function

$$y = x^3$$

gives positive values of dy/dx for all values of x except $x = 0$ where dy/dx is zero. Clearly from Fig. 6.4 it is seen that there is no maximum or minimum at $x = 0$.

6.5   HIGHER ORDER DERIVATIVES

If y is a function of x then dy/dx will in general be also a differentiable function of x. The result of differentiating dy/dx is called the 'second differential coefficient' or 'second derivative'. If the second derivative can be differentiated we obtain the 'third derivative' and so on.

There are standard ways of expressing first and higher derivatives. They may be denoted by

$$\frac{dy}{dx}, \frac{d^2 y}{dx^2}, \frac{d^3 y}{dx^3}, \ldots \frac{d^n y}{dx^n}.$$

Alternative forms are

$$\frac{d}{dx} \cdot y, \left(\frac{d}{dx}\right)^2 y, \left(\frac{d}{dx}\right)^3 y \ldots \left(\frac{d}{dx}\right)^n y;$$

or

$$Dy, D^2 y, D^3 y, \ldots D^n y;$$

or

$$y', y'', y''', \ldots y^{(n)};$$

or, if $y = f(x)$,

$$f'(x), f''(x), f'''(x), \ldots f^{(n)}(x).$$

As an example let us consider the function

$$y = x^5 + 3x^3 + x. \tag{6.1}$$

We differentiate this with respect to x to obtain

$$\frac{dy}{dx} = 5x^4 + 9x^2 + 1.$$

Clearly dy/dx is a function of x which can be differentiated and so we may write

$$\frac{d^2y}{dx^2} = 20x^3 + 18x.$$

We can continue to differentiate and obtain

$$\frac{d^3y}{dx^3} = 60x^2 + 18,$$

$$\frac{d^4y}{dx^4} = 120x,$$

$$\frac{d^5y}{dx^5} = 120, \text{ and } \frac{d^6y}{dx^6} = 0.$$

In this particular example all derivatives higher than the fifth derivative for the function (6.1) vanish.

Some functions can be differentiated indefinitely, for example if we take

$$y = \exp(x)$$

we may write

$$\frac{dy}{dx} = e^x; \frac{d^2y}{dx^2} = e^x; \frac{d^3y}{dx^3}; \ldots \frac{d^ny}{dx^n} = e^x.$$

For the present let us only consider further the second derivative because it is useful in determining maxima and minima. In the previous section we saw that at the region of a maximum dy/dx decreased. If dy/dx is decreasing then d^2y/dx^2 is negative. Likewise at a minimum dy/dx is increasing so d^2y/dx^2 is positive. Thus if we want to be able to distinguish a maximum from a minimum we need only calculate the appropriate values of the second derivative for the values of x at which the first derivative is zero.

As an example let us consider again the function

$$y = x^3 - 6x^2 + 9x + 16$$

which is drawn in Fig. 6.2.

We differentiate to obtain

$$\frac{dy}{dx} = 3x^2 - 12x + 9$$

Maxima or minima occur where $dy/dx = 0$ so we write

$$3x^2 - 12x + 9 = 0$$

or

$$x^2 - 4x + 3 = 0$$

This is rearranged to give

$$(x - 3)(x - 1) = 0$$

and we obtain $x = 3$ and $x = 1$ as values of x where $dy/dx = 0$.

To distinguish between maxima and minima at these two points we obtain the second derivative

$$\frac{d^2 y}{dx^2} = 6x - 12.$$

When $x = 3$,
$$\frac{d^2 y}{dx^3} = 6.$$

The Second derivative is positive when $x = 3$ and this indicates the position of a minimum for the function.

When $x = 1$,
$$\frac{d^2 y}{dx^2} = -6.$$

The second derivative is negative when $x = 1$ and this indicates the position of a maximum for the function.

Exercises 6.5

(1) Determine the first and second derivatives of the following functions

(a) $y = 3 \sin x$

(b) $y = \exp(-4x)$

(c) $y = (x^2 + 1)/x$

(d) $y = x \exp(-x)$

(e) $y = x^2(3 - x)$

(f) $y = \dfrac{1}{1 - x^2}$

(g) $y = \sin \theta - \cos \theta$

(h) $y = \exp(3x)$.

(2) Determine the positions of maxima and minima for the following functions

(a) $y = 16 - 6x - 3x^2$ (b) $y = (x^2 - 4)^2$

(c) $y = \exp(x) - \exp(2x)$ (d) $y = \ln(x) - x$

(3) By using the methods described in this section verify your answers to the problems of Exercises 6.2.

6.6 PROBLEMS ON MAXIMA AND MINIMA

The following simple problems have been chosen to illustrate the usefulness of what we have learned about maxima and minima. The most important step is to set up the appropriate equation for differentiation and this may involve careful algebraic manipulation especially if, at first sight, the equation has several variables. In this case simultaneous equations must be set up so that all the variables except one are eliminated.

Example I

What is the shortest possible diagonal of a rectangle that has an area of 144 sq. in?

    Let x = the length of the rectangle,
       y = the breadth of the rectangle,
       d = the length of the diagonal.

We may immediately write

$$d = (x^2 + y^2)^{1/2}. \tag{6.2}$$

This expresses d as the function of two variables. In order to eliminate one of these variables we set up another equation. We are given that the area of the rectangle is 144 sq. in, so we may write

$$xy = 144 \tag{6.3}$$

or

$$y = \frac{144}{x}$$

We combine equations (6.2) and (6.3) to obtain

$$d = \left\{ x^2 + \frac{144^2}{x^2} \right\}^{1/2}$$

The next step could be the differentiation of d to determine its minimum value but we can make the problem easier by putting

$$z = d^2 = x^2 + \frac{144^2}{x^2}$$

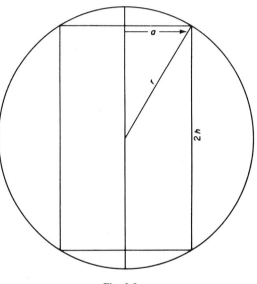

Fig. 6.5.

and then find the minimum value of z because clearly this will give the square of the minimum value of d. We differentiate z and obtain:

$$\frac{dz}{dx} = 2x - \frac{2(144)^2}{x^3} \tag{6.4}$$

By putting dz/dx = 0 for a maximum or a minimum we obtain

$$x^4 = (144)^2$$

i.e.

$$x = 12.$$

We now take the second derivative to check that it is positive when $x = 12$ so that the diagonal is minimum.

For (6.4) we obtain

$$\frac{d^2 z}{dx^2} = 2 + \frac{6(144)^2}{x^4}$$

This is clearly positive whatever the value of x and thus $x = 12$ gives a minimum value of z. We insert this value of x into equation (6.3) to obtain $y = 12$ and these two values enable us to calculate the minimum value of d from expression 6.2:

$$d_{min} = (12^2 + 12^2)^{1/2}$$
$$= 12\sqrt{2}.$$

The minimum length of the diagonal is $12\sqrt{2}$ and this occurs when the length and breadth of the rectangle are each 12 inches long.

Example II

A cylinder is inscribed in a sphere of radius r. What is the maximum volume for such a cylinder.

Let the cylinder have a volume V, height $2h$ and radius a. The sphere and cylinder are shown in cross-section in Fig. 6.5. The volume of the cylinder depends upon the height and the radius

$$V = 2\pi a^2 h.$$

According to the geometry of the system we can write the following equation to relate a and h with the radius of the sphere:

$$r^2 = a^2 + h^2.$$

We may now combine the two equations and obtain

$$V = 2\pi(r^2 - h^2) h.$$

For a maximum or minimum

$$\frac{dV}{dh} = 2\pi r^2 - 6\pi h^2$$
$$= 0,$$

and

$$h = r/\sqrt{3}.$$

The second derivative

$$\frac{d^2 V}{dh^2} = -12\pi h$$

gives a negative value for all positive values of h and thus $h = r/\sqrt{3}$ gives a maximum value for the volume.

$$V_{max} = 2\pi \left(r^2 - \frac{r^2}{3} \right) \frac{r}{\sqrt{3}}$$

$$= \frac{4\pi r^3}{3\sqrt{3}}.$$

Example III

An animal is fed on a diet in which the concentration (F) of an added factor is varied. It is found that the daily consumption (D) of the diet is related to the concentration of the added factor according to the relation

$$D = A - aF \qquad\qquad (6.5)$$

where A is the consumption in the absence of the factor and a is a constant. What value of F must be chosen so that the animal's consumption of the added factor is a maximum?

First of all we can say that the total daily consumption (y) of the factor is given by

$$y = DF.$$

But from (6.5) we know that

$$F = \frac{A - D}{a}$$

so

$$y = \frac{D(A - D)}{a}$$

For a maximum or a minimum

$$\frac{dy}{dD} = \frac{1}{a}(A - 2D)$$

$$= 0,$$

i.e. $A = 2D$.

Second differentiation shows that this gives a maximum value of F and therefore

$$F = \frac{A}{2a}.$$

Exercises 6.6

(1) A lidless box has a square base and vertical sides. Assuming that there is no overlap at the joints, what is the minimum area of material that is needed to make the box contain a volume V? (The answer should be a formula relating V to the base length and height of the box).

(2) A piece of wire of length L is cut into two pieces. One piece is bent to form a circle, the other an equilateral triangle. What are the respective lengths of the parts when the sum of the areas is minimum?

(3) A lion is running due west at a speed of 20 m.p.h. It passes a point one quarter of a mile due south of a hunter. The hunter wishes to photograph the lion but his car has a maximum speed of 18 m.p.h. What direction must the hunter go to get as close to the lion as possible and how close will he then be to the lion?

(4) Suppose that the rate of energy utilisation of a fish swimming along a river varies as the cube of its speed through the water. Show that the most economical speed against a current will be equal to $1\frac{1}{2}$ times the current.

(5) The velocity (v) of a certain class of chemical reactions obeys the relationship

$$v = k(b + x)(c - x)$$

where k, b and c are positive constants and x is the amount of substrate that has decomposed. Show that v is maximum when $x = \frac{1}{2}(c - b)$.

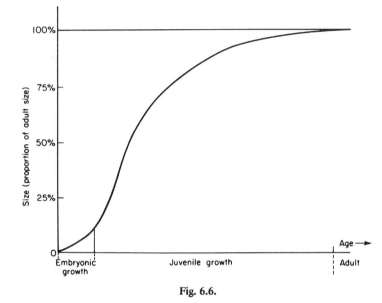

Fig. 6.6.

6.7   GROWTH CURVES

If an organism, tissue or culture of bacteria had a constant rate of growth in the sense that each cell divides after a period of time τ then, as we have seen in previous chapters, its size would increase exponentially with time. This pattern of growth is rarely achieved in practice except perhaps in the initial stages. Generally, there are limitations upon the ultimate size that may be attained and the graph of mass (x) over a period of time will have a shape similar to that shown in Fig. 6.6. This curve is characterised by the fact that initially the curve has an exponential form but this is not maintained and the curve plateau's as the adult size is attained.

Attempts have been made to produce formulae to fit the growth curves. Although none has been found which accurately fits all the possible cases there are one or two which, by a suitable choice of constants, can give a reasonable fit in most cases. We will consider one of these curves *The Logistic Growth Curve*:

This has the form:

$$x = \frac{a}{(1 + b \exp[-kt])} \tag{6.6}$$

where x is the size at time t and the positive constants a, b and k are characteristic of the organism, tissue or culture that is being studied.

The rate of growth v is obtained by differentiating (6.6) with respect to t:

$$v = \frac{dx}{dt} = \frac{abk\,e^{-kt}}{(1 + b\exp[-kt])^2} = \frac{x^2}{a}\,bk\,e^{-kt}. \tag{6.7}$$

In order to determine the time at which the maximum rate of growth takes place we differentiate v:

$$\frac{dv}{dt} = \frac{2bk}{a}\,x\,\frac{dx}{dt}\,e^{-kt} - \frac{bk^2}{a}\,x^2 e^{-kt}$$

Expression (6.7) for dx/dt is inserted into the equation to give:

$$\frac{dv}{dt} = \frac{2x^3}{a^2}\,b^2 k^2 \exp(-2kt) - \frac{x^2}{a}\,bk^2 \exp(-kt)$$

$$= \frac{bk^2}{a}\,x^2 \exp(-kt)\left\{\frac{2xb}{a}\exp(-kt) - 1\right\}.$$

For a maximum or minimum value

$$\frac{dv}{dt} = 0$$

in which case

$$x = 0,$$

or

$$\exp(-kt) = 0$$

or

$$\left\{\frac{2xb}{a}\exp(-kt) - 1\right\} = 0.$$

If $x = 0$ then the denominator of expression (6.6) must be infinite since the numerator a is a constant. The denominator is infinite if $t = -\infty$ and therefore a maximum or a minimum for this value of t is of no interest to us.

If $\exp(-kt) = 0$, then $t = \infty$ and again this is of no interest to us.

If

$$\left(\frac{2xb}{a}\exp[-kt] - 1\right) = 0 \tag{6.8}$$

then substituting the expression for x into (6.8) gives

$$\frac{2b\,e^{-kt}}{1 + b\,e^{-kt}} - 1 = 0.$$

Rearrangement of this equation gives

$$2b\,e^{-kt} = 1 + b\,e^{-kt}$$

or

$$e^{-kt} = \frac{1}{b}$$

$$t = \frac{1}{k}\ln b. \tag{6.9}$$

A maximum or a minimum value of v occurs at a point in time given by expression (6.9). The value of x at this point is given by inserting expression (6.9) into expression (6.6) and we obtain

$$x = a/2 \tag{6.10}$$

The value of v at this time is obtained from (6.7) and is given by

$$v = ak/4.$$

In order to determine if this gives a maximum value of v we take the expression for dv/dt and differentiate it again:

$$\frac{d^2 v}{dt^2} = \left[6x^2 \cdot \frac{dx}{dt}\left(\frac{bk}{a}\right)^2 \cdot e^{-2kt} \right] - \left[4x^3 \left(\frac{bk}{a}\right)^2 \cdot k\,e^{-kt} \right]$$

$$- \frac{2x}{a}\frac{dx}{dt}(bk^2)\,e^{-kt} + \frac{x^2}{a}\,bk^3\,e^{-kt}$$

$$= 6x^2 \left(\frac{x^2}{a}\,bk\,e^{-3kt}\right)\left(\frac{bk}{a}\right)^2 - 4kx^3 \left(\frac{bk}{a}\right)^2 \cdot e^{-2kt}$$

$$- \frac{2x^3}{a^2}\,b^2 k^3\,e^{-2kt} + \frac{x^2}{a}\,bk^3\,e^{-kt}.$$

For

$$x = a/2$$

and

$$t = \frac{1}{k} \ln b$$

we obtain

$$\frac{d^2 v}{dt^2} = \tfrac{3}{8} ak^3 - \tfrac{1}{2} ak^3 - \tfrac{1}{4} ak^3 + \tfrac{1}{4} ak^3$$

$$= -\tfrac{1}{8} ak^3.$$

The negative value indicates that a maximum value for the rate of growth occurs at time

$$t = \frac{1}{k} \ln b.$$

6.8   POSITION, VELOCITY AND ACCELERATION

Consider a point moving along a straight line and that its position Y with respect to a fixed point O (Fig. 6.7) is measurable, the distance OY denoted by y. In the figure the distance OY is reckoned positive, but if the point takes a position such as Y′ on the other side of O along the line, its distance from O

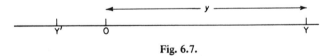

Fig. 6.7.

is reckoned to be negative. Since the point is moving along the line, y will have different values for different values of time t, i.e. y will be a function of t and the relationship between them may be expressed as an equation and plotted graphically. For example

$$y = t^4 - t^2.$$

Such an expression relating position and time is known as a *law of motion*, and we can see from this particular expression and from its graphical representation in Fig. 6.8 that the point will be at O when $t = 0$ and 1. (It will also be at O when $t = -1$ but we are only interested in positive values of t in the present discussion.)

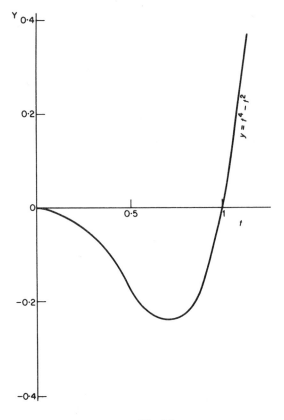

Fig. 6.8.

The velocity v of the point along the line, i.e. the rate of change of position, can be calculated by differentiation, and for our example we have:

$$v = \frac{dy}{dt} = 4t^3 - 2t.$$

The velocity is reckoned positive when the point is moving in the direction $\overrightarrow{Y'OY}$.

The acceleration f is defined as the rate of change of velocity and is calculate by differentiating the velocity

$$f = \frac{dv}{dt} = \frac{d^2y}{dt^2} = 12t^2 - 2$$

If we wish to calculate the value of the maximum or minimum value of the velocity we put

$$\frac{dv}{dt} = 0$$

Since differentiation of the velocity is equal to the acceleration, it follows that in order to have a maximum or minimum value of the velocity, the acceleration must be zero.

For our particular example

$$\frac{dv}{dt} = 12t^2 - 2 = 0$$

This gives a maximum or minimum value of v when

$$t = \frac{1}{\sqrt{6}} .$$

Differentiation of f gives

$$\frac{df}{dt} = 24t$$

This is positive when $t = 1/\sqrt{6}$ and a minimum value is obtained at this point.

Exercises 6.8

(1) Derive expressions for the velocity (v) and acceleration (f) for the following laws of motion:

(a) $y = A \exp(-kt)$ (b) $y = A \ln 3t$

(c) $y = 4t + 3$ (d) $y = ut + \frac{1}{2} gt^2$

(2) Calculate the minimum values of the velocity y of the points which obey the following laws of motion:

(a) $y = t^3 - 3t$ (b) $y = A(t^3 - 3t^2)$.

7

INTEGRATION

7.1 INTRODUCTION

In previous chapters we showed how to find the differential coefficient of a function with respect to its independent variable. For instance, given

$$y = f(x) = 2x^3 + 3x^2 + x,$$

we learned how to evaluate its differential coefficient dy/dx and obtain

$$\frac{dy}{dx} = 6x^2 + 6x + 1.$$

Now suppose that we are presented with the reverse problem, we are told what the differential coefficient of the function is and asked to find the function that has been differentiated. Thus for the above quoted example, if instead of being given

$$f(x) = 2x^3 + 3x^2 + x$$

and asked to find dy/dx, we are given that

$$\frac{df(x)}{dx} = 6x^2 + 6x + 1$$

and asked to find the function $f(x)$.

The process of finding the function for which the differential coefficient is known is called *integration*. The function that is integrated is known as the *integrand*.

Briefly, we may state that for

$$y = f(x),$$

we differentiate $f(x)$ to obtain dy/dx, and we integrate dy/dx to obtain $f(x)$. The function obtained as a result of integration is known as an *integral.*

In this chapter we consider some of the methods by which the process of integration is carried out. From the point of view of the biologist, this chapter may not be very satisfying because no examples of a biological nature can be quoted until quite a fair amount of the chapter has been covered. This is due to the nature of the topic that is being covered; quite a bit on integration has to be discussed before anything of direct relevance can be introduced.

7.2 DEFINITION

The integral of a function $f(x)$ with respect to x is the function whose differential coefficient with respect to x is $f(x)$, and it is written $\int f(x)dx$.

The symbol $\int \ldots dx$ in words is 'the integral of . . . with respect to x'. The symbol $\int$ is never used by itself; it always appears in the combination $\int \ldots dx$. The variable need not always be denoted by x, so sometimes we will come across such expressions as

$$\int f(u)\, du; \int g(t)\, dt \text{ etc.}$$

Thus

$$\int u^2\, du$$

stands for the integral of u^2 with respect to u;

$$\int \ln (at)\, dt$$

stands for the integral of $\ln (at)$ with respect to t.

To take a simple case we know that the differential coefficient of x^2 is $2x$, so that from the definition of an integral we integrate $2x$ to obtain x^2. In other words if

$$y = x^2$$

then

$$\frac{dy}{dx} = 2x$$

and

$$\int 2x\, dx = x^2.$$

Unfortunately, since the methods of integration consist of the retracing of the steps of differentiation there is no general method which may be applied. The evaluation of a differential coefficient is a direct operation for which precise rules exist to obtain a unique result. No such rules exist for integration. The function to be integrated must be recognisable as the result of a differentiation of some kind.

A point to notice is that the definition of an integral given at the beginning of this section is such that the integral of a function can not be uniquely found because the differential coefficient of a constant is zero. Therefore the integration of zero will give a constant whose value is unspecified. To clarify this point let us consider again the example.

$$\int 2x \, dx = x^2.$$

This is not a unique solution because we know that if

$$y = x^2 + C, \text{ where } C \text{ is a constant, then}$$

$$\frac{dy}{dx} = 2x.$$

Therefore, the integration of $2x$ with respect to x gives the result

$$\int 2x \, dx = x^2 + C$$

and C is a constant which may take any value
    Generally, we state

$$\int g(x) \, dx = f(x) + C \qquad\qquad (7.1)$$

where C is an arbitrary constant and $g(x)$ is the differential coefficient of $f(x)$. Since the integration yields a function which contains an arbitrary constant such integrals are known as *indefinite integrals.*

7.3   THE INTEGRAL OF $x^n \, (n \neq -1)$

From the chapter on differentiation we know that

$$\frac{d(x^n)}{dx} = nx^{n-1}.$$

We can replace n by $n + 1$ and obtain

$$\frac{d(x^{n+1})}{dx} = (n + 1) x^n.$$

This can be rearranged to take the form

$$\frac{1}{n+1} \frac{d(x^{n+1})}{dx} = x^n, \qquad\qquad (7.2)$$

or

$$\frac{d}{dx}\left(\frac{x^{n+1}}{n+1}\right) = x^n.$$

Therefore, by the definition of an integral, we may write

$$\int x^n \, dx = \frac{x^{n+1}}{n+1} + C. \tag{7.3}$$

This result holds for all values of n except for $n = -1$.

Example I

$$\int x^3 \, dx.$$

In this case $n = 3$ which, when inserted into the formula of (7.3) gives

$$\int x^3 \, dx = \frac{x^4}{4} + C.$$

Example II

$$\int x^{-1/2} \, dx$$

In this case $n = -\frac{1}{2}$ so that

$$\int x^{-1/2} \, dx = \frac{x^{-1/2 + 1}}{-\frac{1}{2}+1} + C = -2x^{1/2} + C.$$

Exercises 7.3

Evaluate the following indefinite integrals:

(1) $\displaystyle\int x^{-5} \, dx$ (2) $\displaystyle\int x^{-0.1} \, dx$

(3) $\displaystyle\int dx$ (4) $\displaystyle\int x^{2.5} \, dx.$

7.4   THE INTEGRAL OF x^{-1}

Let us consider again the relationship

$$\frac{d(x^n)}{dx} = nx^{n-1}.$$

If we put $n = 0$, we obtain

$$\frac{d(x^0)}{dx} = 0 \cdot x^{-1} = 0.$$

We cannot then use the relationship to give us the integral of x^{-1}. However, we know from Section 5.14 that

$$\frac{d}{dx}(\ln x) = x^{-1},$$

therefore, we may state

$$\int x^{-1}\, dx = \ln x + C \qquad\qquad (7.4)$$

We may however slightly alter the above expression by putting

$$\ln c = C.$$

Where c is a constant, in which case we have

$$\int x^{-1}\, dx = \ln x + \ln c$$
$$= \ln (cx).$$

7.5   INTEGRATION OF TRIGONOMETRIC FUNCTIONS

Only two functions will be considered at this stage. In Chapter 5 on differentiation we were told that

$$\frac{d}{dx}(\sin x) = \cos x,$$

therefore

$$\int \cos x\, dx = \sin x + C, \qquad\qquad (7.5)$$

and

$$\frac{d}{dx}(\cos x) = -\sin x,$$

therefore

$$\int \sin x \, dx = -\cos x + C.$$

(7.6)

7.6   INTEGRATION OF e^x

Integration of e^x is very easy because we know that

$$\frac{d}{dx}(\exp x) = \exp x,$$

therefore

$$\int \exp x \, dx = \exp x + C.$$

(7.7)

7.7   INTEGRATION OF $af(x)$ (a = constant)

We have shown in previous sections how the indefinite integral for the simple mathematical functions may be obtained. In the following sections we consider some rules that are available for integration of functions which are sums, multiples and products etc. of these simpler functions. We consider first of all the integration of any function multiplied by a constant. In other words, if we know how to integrate $f(x)$ we want to know also how to integrate $af(x)$ where a is a constant.

Suppose that

$$\int f(x) \, dx = g(x) + C,$$

then

$$\frac{d}{dx}\{g(x)\} = f(x).$$

We can multiply both these expressions throughout by the constant a to obtain

$$a \int f(x) \, dx = ag(x) + C$$

(7.8)

and

$$a \frac{d}{dx}[g(x)] = a f(x)$$

(7.9)

Note that we do not need to multiply the constant C by a to obtain aC because C represents an arbitrary constant having any value.

According to the result of Section 5.8 we may write

$$a \frac{d}{dx} [g(x)] = \frac{d}{dx} [a g(x)]$$

and we use this to put expression (7.9) in the form

$$\frac{d}{dx} [a g(x)] = a f(x).$$

From the definition of an indefinite integral we may write

$$\int a f(x) \, dx = a g(x) + C.$$

If we now compare this expression with expression (7.8) we see that

$$\int a f(x) \, dx = a \int f(x) \, dx. \qquad (7.10)$$

In words this result means that if we have to integrate a function multiplied by a constant we can integrate the function and then multiply the answer by a constant.

Example I

Find the indefinite integral

$$\int 2x^2 \, dx.$$

From expression (7.10), by putting $a = 2$ and $f(x) = x^2$, we may write

$$\int 2x^2 \, dx = 2 \int x^2 \, dx.$$

From the result of Section 7.3 we may write

$$\int x^2 \, dx = \frac{x^3}{3} + C,$$

and therefore

$$2 \int x^2 \, dx = \tfrac{2}{3} x^3 + C.$$

Exercises 7.7

Find the following indefinite integrals:

(1) $\int 4 \cos (x)\, dx$

(2) $\int 6(x)^{1/2}\, dx$

(3) $\int 8x^2\, dx$

(4) $\int \dfrac{1}{(8x^2)^{1/3}}\, dx.$

7.8   TO INTEGRATE THE SUM OR DIFFERENCE OF TWO FUNCTIONS

Suppose that we are able to integrate two simple functions: $f(x)$ and $g(x)$, and that the indefinite integrals are expressed as follows:

$$\int f(x)\, dx = F(x) + C$$

and   (7.11)

$$\int g(x)\, dx = G(x) + C.$$

We want to be able to perform the integration of the sum or the difference of the two functions, i.e. to evaluate

$$\int [f(x) \pm g(x)]\, dx$$

and to obtain the answer in terms of $F(x)$ and $G(x)$.

From expressions (7.11) we may write

$$\frac{d}{dx} F(x) = f(x)$$

(7.12)

$$\frac{d}{dx} G(x) = g(x).$$

Also we may add or subtract the expressions (7.11) to obtain

$$\int f(x)\, dx \pm \int g(x)\, dx = [F(x) \pm G(x)] + C.$$ (7.13)

Note that we do not put $2C$ in this expression as C is an arbitrary constant having any value.

Likewise we may add or subtract expressions (7.12) to obtain;

$$\frac{d}{dx} F(x) \pm \frac{d}{dx} G(x) = [f(x) \pm g(x)],$$ (7.14)

but according to the result of Section 5.6 we may write

$$\frac{d}{dx} F(x) \pm \frac{d}{dx} G(x) = \frac{d}{dx} [F(x) \pm G(x)]$$

so that expression (7.14) takes the form

$$\frac{d}{dx} [F(x) \pm G(x)] = [f(x) \pm g(x)].$$

From this expression we obtain, according to the definition of an integral

$$\int [f(x) \pm g(x)] \, dx = [F(x) \pm G(x)] + C.$$

If we compare this expression with expression 7.13 we obtain the result

$$\int [f(x) \pm g(x)] \, dx = \int f(x) \, dx \pm \int g(x) \, dx. \qquad (7.15)$$

In words this result is stated simply as the integral of the sum or difference of two functions is equal to the sum or difference of the integrals of the two functions. This result is now illustrated by way of examples.

Example I

Evaluate the indefinite integral

$$\int [x^2 + \cos x] \, dx.$$

From our result we may immediately write

$$\int [x^2 + \cos x] \, dx = \int x^2 \, dx + \int \cos x \, dx.$$

From the results of previous sections we may write

$$\int x^2 \, dx = \frac{x^3}{3} + C.$$

and

$$\int \cos x \, dx = \sin x + C.$$

We now add the two results to obtain:

$$\int [x^2 + \cos x] \, dx = \frac{x^3}{3} + \sin x + C.$$

It should be obvious at this stage that the result can be extended to the integration of the sums or differences of more than two functions. Thus we may write

$$\int [f(x) \pm g(x) \pm h(x) \pm j(x) \text{ etc.}]\,dx$$

$$= \int f(x)\,dx \pm \int g(x)\,dx \pm \int h(x)\,dx \pm \int j(x)\,dx \text{ etc.}$$

(7.16)

Exercises 7.8

(1) Expand e^x according to the definition contained in Section 2.2 and use the expansion to prove that

$$\int e^x\,dx = e^x + C.$$

7.9   TO DETERMINE THE INDEFINITE INTEGRALS OF THE FORM

$$\int f(ax + b)\,dx.$$

So far we have learned how to determine the indefinite integrals illustrated by the following examples:

$$\int \cos x\,dx,$$
$$\int \exp[x]\,dx,$$
$$\int x^4\,dx.$$

We now wish to use these integrals to evaluate integrals of the form

$$\int \cos(6x + \pi)\,dx,$$
$$\int \exp[10x + 4]\,dx,$$
$$\int (\tfrac{1}{2}x + 1)^4\,dx.$$

In other words we want to be able to find

$$\int f(ax + b)\,dx$$

by using the known result of

$$\int f(x)\,dx.$$

First of all let us write

$$u = ax + b$$

so that

$$\frac{du}{dx} = a. \tag{7.17}$$

Also let us write

$$\int f(ax + b)\, dx = \int f(u)\, dx = F(u), \tag{7.18}$$

from which we may obtain

$$\frac{dF(u)}{dx} = f(u) \tag{7.19}$$

We now use one of the results from the chapters on differentiation to obtain

$$\frac{dF(u)}{dx} = \frac{dF(u)}{du} \cdot \frac{du}{dx},$$

but according to expression (7.17) this may be simplified to the form

$$a\,\frac{dF(u)}{du} = \frac{dF(u)}{dx}$$

Further, according to expression (7.19) we may write

$$a\,\frac{dF(u)}{du} = f(u)$$

or

$$\frac{dF(u)}{du} = \frac{1}{a}\, f(u).$$

According to the definition of the indefinite integral we may now write

$$F(u) = \int \frac{1}{a}\, f(u)\, du = \frac{1}{a} \int f(u)\, du.$$

If we compare this expression with expression (7.18) we obtain the result

$$\int f(u)\, dx = \frac{1}{a} \int f(u)\, du. \tag{7.20}$$

We now use this result to solve the integrals given in the following examples:

Example I

To evaluate $\int \cos{(6x + \pi)}\, dx.$

First of all put

$$u = (6x + \pi),$$
$$a = 6,$$
$$b = \pi$$

and

$$f(u) = \cos u.$$

Next we make use of expression (7.20) and write

$$\int \cos (6x + \pi) \, dx = \int \cos u \, dx = \tfrac{1}{6} \int \cos u \, du.$$

But we know that

$$\int \cos u \, du = \sin u + C,$$

therefore, we may obtain the final answer,

$$\int \cos (6x + \pi) \, dx = \tfrac{1}{6} [\sin (6x + \pi) + C].$$

If you are in any doubt about the answer it may be easily verified by differentiating the result. Indeed, it is usually good practice to check all integrations in this way.

Example II

To evaluate $\int \exp[10x + 4] \, dx.$

In this case we put

$$u = 10x + 4,$$
$$a = 10,$$
$$b = 4,$$
$$f(u) = \exp[u].$$

From expression (7.20) we can write

$$\int \exp[10x + 4] \, dx = \tfrac{1}{10} \int \exp[u] \, du$$
$$= \tfrac{1}{10} \exp[u] + C$$
$$= \tfrac{1}{10} \exp[10x + 4] + C.$$

Example III

To evaluate $\int (\tfrac{1}{2}x + 1)^4\, dx$

put
$$u = (\tfrac{1}{2}x + 1),$$
$$a = \tfrac{1}{2},$$
$$b = 1,$$
$$f(u) = u^4.$$

From expression (7.20) we write

$$\int (\tfrac{1}{2}x + 1)^4\, dx = \int u^4\, dx = 2 \int u^4\, du$$
$$= \tfrac{2}{5} u^5 + C = \tfrac{2}{5}(\tfrac{1}{2}x + 1)^5 + C.$$

Exercises 7.9

Evaluate the following indefinite integrals and check your answers by differentiation:

(1) $\int \dfrac{dx}{(ax + b)}$ a and b are constants   (2) $\int (3x + 2)^5\, dx$

(3) $\int \sin (6t - \tfrac{1}{2}\pi)\, dt$ (4) $\int (3 - x)^{1/2}\, dx$.

(5) $\int (t + 3)^{-3}\, dt$

7.10   AREAS UNDER CURVES

Although there is much more to be discussed about the indefinite integral we are ready at this stage to make use of what we already know and in this section we consider how the processes of integration can be used to calculate exactly the area under a curve.

Consider the curve shown in Fig. 7.1 which is a representation of an arbitrary curve denoted by

$$y = f(x).$$

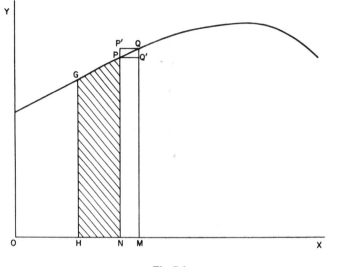

Fig. 7.1.

Let GP be an arc of the curve measured from some fixed point G to a variable point P with coordinates (x, y). Let the area of the figure GHNP, that is the area between the curve, the X-axis and the ordinates GH and PN be denoted by A. Now P is a variable point so consequently if P is altered then the area of GHNP will alter. To each value of the abscissa x there will be a corresponding value of the area A. Thus the area of GHNP is some function of x which may be represented

$$A = F(x).$$

We now choose a point Q on the curve which is near to P so that the co-ordinates of Q are $(x + \delta x, y + \delta y)$. In Fig. 7.1 the ordinate of Q is MQ so that

$$MQ = y + \delta y$$

and

$$OM = x + \delta x.$$

The area of GHMQ is larger than the area of GHNP by an amount which may be represented by δA, i.e.

$$\text{area GHMQ} = A + \delta A.$$

The lines P'Q and PQ' are now drawn parallel to the X-axis to complete the rectangle P'QQ'P. Thus from the figure the area of PQMN is equal to δA.

Also from the figure it is seen that the area of the rectangle NMQP′ is greater than δA and the area of the rectangle NMQ′P is less than δA. But

$$\text{area NMQP}' = \text{NM . MQ} = \delta x(y + \delta y)$$

and

$$\text{area NMQ}'\text{P} = \text{NM . MQ}' = \delta x . y,$$

so that

$$\delta A < (y + \delta y)\, \delta x$$

and

$$\delta A > y . \delta x,$$

where the symbols $<$ and $>$ mean 'less than' and 'greater than' respectively.

The above two relationships, which are known as inequalities, may be combined together in the following way:

$$(y + \delta y)\, \delta x > \delta A > y . \delta x.$$

We can divide this throughout by δx to obtain

$$(y + \delta y) > \frac{\delta A}{\delta x} > y.$$

We note from this relationship that $\delta A/\delta x$ differs from y by an amount smaller than δy. Also, as δx tends to zero so δy tends to zero in which case $\delta A/\delta x$ will tend to the limiting value of y. We may express this last statement in the form

$$\underset{\delta x \to 0}{\text{Lim}}\ \frac{\delta A}{\delta x} = y.$$

However, we know from the definition of the differential coefficient that

$$\underset{\delta x \to 0}{\text{Lim}}\ \frac{\delta A}{\delta x} = \frac{\mathrm{d}A}{\mathrm{d}x}.$$

We compare these last two expressions and obtain the relationship

$$\frac{\mathrm{d}A}{\mathrm{d}x} = y = \mathrm{f}(x). \tag{7.21}$$

Since both A and y are functions of x, we can use the definition of the indefinite integral to obtain the relationship

$$A(x) = \int y \, dx = \int f(x) \, dx. \qquad (7.22)$$

Now at the beginning of the section we use $A(x)$ to denote the area of the figure GHNP. At first sight then it seems that we have only to evaluate the indefinite integral to obtain the expression for the required area. However the problem does arise that the integration will involve an arbitrary constant which can take any value. Our next step is to determine just one value for this constant which will give us the correct value for the area. Let us take another look at Fig. 7.1. We specified the required area as being to the right of the ordinate GH. Let us now consider the case where the area gets smaller and smaller because we decrease the value of x so that N becomes coincident with H. In this case the area is zero and this fact enables us to calculate the value of C as follows:

Assign to the point G the co-ordinates (x', y') so that

$$OH = x'.$$

Also from expression (7.22) we may write

$$A(x) = \int y \, dx = F(x) + C \qquad (7.23)$$

where $F(x)$ is obtained by performing the integration. Now we know that when $x = x'$ the area is zero so that we may write:

$$A(x') = F(x') + C = 0. \qquad (7.24)$$

Thus we obtain a value of $C = -F(x')$ which gives us the correct value of the area. We use expression (7.24) to modify expression (7.23) to take the form

$$A(x) = F(x) - F(x') \qquad (7.25)$$

This yields a relationship for the value of the area of GHNP which lies within the arc GP, the X-axis and the ordinates GH and PN. If now we fix a definite value to the co-ordinates of P say (x'', y''), then the area GHNP is given by

$$\text{area GHNP} = F(x'') - F(x') \qquad (7.26)$$

This result is now used to determine the areas according to the following examples:

Example 1

Find the area contained by the ordinates $x = 1$, and $x = 2$, the X-axis and the curve $y = x^2$.

First of all we draw a rough sketch of the curve (Fig. 7.2). Let A and B respectively be the points on the curve whose abscissae are 1 and 2. We want to determine

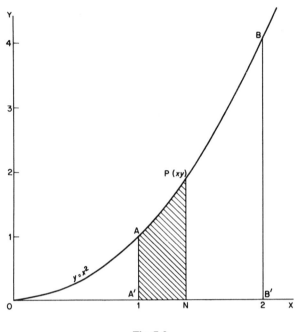

Fig. 7.2.

the area AA′B′B. Let P(xy) be any point lying on the curve between A and B. From expression (7.23) above we write

$$\text{Area AA'NP} = \int y \, dx.$$

For our particular case we write

$$\text{Area AA'NP} = \int x^2 \, dx$$

$$= \frac{x^2}{3} + C.$$

When $x = 1$, area AA′NP = 0 because N is coincident then with A′. Therefore

$$0 = \tfrac{1}{3} + C$$

and therefore

$$C = -\tfrac{1}{3}.$$

The general expression for the area is thus

$$A(x) = \frac{x^3}{3} - \frac{1}{3}$$

When we insert the particular value $x = 2$ we obtain the required answer

$$\text{Area AA}'\text{B}'\text{B} = \frac{(2)^3}{3} - \frac{1}{3}$$

$$= \tfrac{8}{3} - \tfrac{1}{3}$$

$$= \tfrac{7}{3} .$$

Example II

Find the area bounded by the curve $y = \exp[-3t]$, the t-axis and the ordinates $t = \tfrac{1}{3}$ and $t = \infty$.

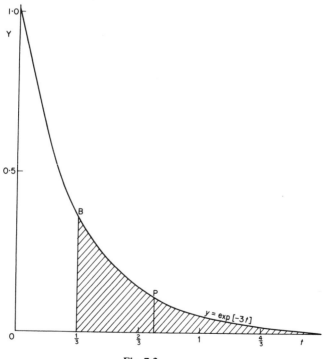

Fig. 7.3.

A rough sketch of the curve is shown in Fig. 7.3. Let B be the point on the curve whose abscissa is $\frac{1}{3}$. For a point P on the curve whose abscissa exceeds $\frac{1}{3}$ we write

$$A(t) = \int y \, dt$$
$$= \int \exp[-3t] \, dt$$
$$= -\tfrac{1}{3} \exp[-3t] + C.$$

When P coincides with B, the area A is zero. We therefore put $t = \frac{1}{3}$ in the above equation and equate to zero

$$0 = -\tfrac{1}{3} \exp[-1] + C.$$

This gives our value for C:

$$C = \tfrac{1}{3} e^{-1} = \frac{1}{3e}$$

The value of the area for any value of t greater than $\frac{1}{3}$ is thus given by

$$A(t) = -\tfrac{1}{3} \exp[-3t] + \frac{1}{3e}.$$

$$A = -\tfrac{1}{3} \exp[-\infty] + \frac{1}{3e}$$

when $t = \infty$

$$= -\frac{1}{3\,e^{\infty}} + \frac{1}{3\,e}$$

$$= 0 + \frac{1}{3\,e}$$

$$= \frac{1}{3\,e}.$$

Example III

Find the area bounded by the curve $y = 1/x$, the X-axis and the ordinates $x = a$ and $x = b$.

The shaded area in the rough sketch (Fig. 7.4) is the area to be determined.

$$A(x) = \int \frac{1}{x} \, dx$$
$$= \ln(x) + C.$$

When $x = a$, $A(x) = 0$ therefore

$$0 = \ln (a) + C,$$

and

$$C = - \ln (a).$$

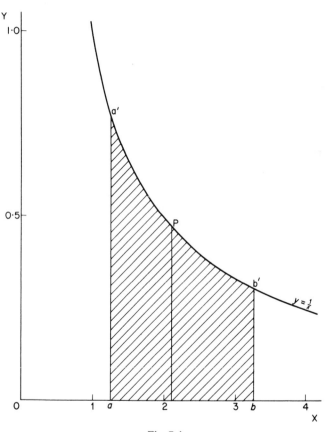

Fig. 7.4.

The general equation for the area is

$$A(x) = \ln (x) - \ln (a).$$

If we put $x = b$ we obtain the value for the required area $aa'b'b$:

$$\text{Area } aa'b'b = \ln (b) - \ln (a)$$

$$= \ln (b/a).$$

Exercises 7.10

Find the areas bounded by the curves, ordinates and X-axis for the following examples:

(1) $y = (3x + 2), x = 0, x = 1.$ (2) $y = (1 - \exp[-3x]), x = 1, x = 3.$

(3) $y = \sin (3x), x = 0, x = \pi/6.$ (4) $y = (x + 3)^{-1}, x = 2, x = 7.$

(5) $y = (3x + 9)^3, x = 0, x = 2.$ (6) $y = (3x - 2)^{-1/2}, x = \frac{2}{3}, x = \frac{18}{3}.$

(7) $y = (x + 1)^{1/2} + (x + 1)^{-1/2}, x = 3, x = 8.$

7.11   THE DEFINITE INTEGRAL

In the previous section we learned how to calculate the area of a figure bounded by the curve of a function, the X-axis and any two ordinates. If the function is denoted $f(x)$ and the two ordinates are given by $x = a$ and $x = b$, then we may use a shorthand method to denote the area A:

$$A = \int_a^b f(x) \, dx. \qquad (7.27)$$

Alternatively we may say that the above expression denotes the area of the curve of $f(x)$ between the limits $x = a$ and $x = b$. In expression (7.27) we meet what is known as a *definite integral;* there is only one area that it denotes and thus there is no arbitrary constant. The ordinate $x = b$ is known as the *upper limit* or *superior limit;* the ordinate $x = a$ is known as the *lower limit* or *inferior limit.*

Now consider the indefinite integral

$$\int f(x) \, dx = F(x) + C. \qquad (7.28)$$

In the previous section we calculated the area between two limits by first of all determining the required value of C. Thus if the lower limit is $x = a$ we write

$$F(a) + C = 0$$

$$C = - F(a).$$

The required area is now found by inserting the value $x = b$:

$$\text{Area} = F(b) + C$$

$$= F(b) - F(a).$$

If we now refer to expression (7.27) we see that

$$\int_a^b f(x)\, dx = F(b) - F(a) \tag{7.29}$$

There is also a shorthand way of denoting $F(b) - F(a)$. It is often written

$$F(a) - F(b) = [F(x)]_a^b \tag{7.30}$$

where $F(x)$ is the indefinite integral of $f(x)$ with the arbitrary constant put equal to zero.

The form (7.30) is usually used when we evaluate definite integrals as in the following examples:

Example I

Evaluate $\int_1^3 x^3\, dx.$

The indefinite integral is

$$\int x^3\, dx = \frac{x^4}{4} + C,$$

so that

$$\int_1^3 x^3\, dx = \left[\frac{x^4}{4}\right]_1^3 = \frac{3^4}{4} - \frac{1}{4}$$

$$= \frac{81}{4} - \frac{1}{4} = \frac{80}{4}$$

$$= 20.$$

Example II

Evaluate

$$\int_1^e \frac{1}{x}\, dx.$$

Since we know how to integrate $1/x$, we can straight away write

$$\int_1^e \frac{1}{x}\, dx = \left[\ln x\right]_1^e = \ln(e) - \ln(1)$$

$$= 1 - 0$$

$$= 1.$$

Exercises 7.11

Evaluate the following definite integrals:

(1) $\int_{1}^{5} (2x + x^2)\, dx$

(2) $\int_{1}^{3} (2x + x^2)\, dx$

(3) $\int_{3}^{5} (2x + x^2)\, dx$

(4) $\int_{0}^{1/2} (1 - x)^{-1/2}\, dx$

(5) $\int_{2}^{2 \cdot 5} \exp[x]\, dx$

(6) $\int_{1}^{6} \dfrac{dx}{(3x - 2)^{1/2}}$

(7) $\int_{1}^{2} \left(3 \exp(2x) + \dfrac{1}{(x + 1)^2}\right) dx$

(8) $\int_{1}^{3} \dfrac{dx}{(4 - x)^3}$

(9) $\int_{\frac{\pi}{2}}^{\pi} \sin(x/2)\, dx$

(10) $\int_{0}^{3} (t + 3)^2\, dt$

(11) $\int_{4}^{5} (x - 3)^{-5}\, dx$

(12) $\int_{2 \cdot 5}^{6} (t + 2)^{1/3}\, dt$

(13) $\int_{4}^{12} (x + \tfrac{1}{2})^{-1/2}\, dx$

(14) $\int_{1/12}^{1/8} \cos[4\pi t - \pi/3]\, dt$

(15) $\int_{0}^{1} (3x + 9)^3\, dx$

(16) $\int_{1}^{10} x^{-4}\, dx$

(17) $\int_{0}^{2} (2 - x)^{1/2}\, dx$

(18) $\int_{0}^{1} (x + 1)^{-1}\, dx$

(19) $\int_{4}^{5} \left(\dfrac{x}{4} - 1\right)^5 dx$

(20) $\int_{4}^{5} (3x - 2)^{-1/2}\, dx$

(21) $\int_{\frac{\pi}{9}}^{\frac{5\pi}{9}} \sin\left[\tfrac{3}{2}x\right] dx$

(22) $\int_{1/2}^{1} (1 - \exp[-2t])\, dt$

(23) Prove that, if $c > b > a$, then

$$\int_{a}^{c} f(x)\, dx = \int_{a}^{b} f(x)\, dx + \int_{b}^{c} f(x)\, dx.$$

7.12 'NEGATIVE' AREAS

Consider the integral

$$\int_{4}^{5}(4-x)\,dx$$

The indefinite integral with arbitrary constant zero is

$$\int(4-x)\,dx = 4x - \frac{x^2}{2}$$

and thus the definite integral is

$$\int_{4}^{5}(4-x)\,dx = \left[4x - \frac{x^2}{2}\right]_{4}^{5}$$
$$= (20 - \tfrac{25}{2}) - (16 - \tfrac{16}{2})$$
$$= -\tfrac{1}{2}.$$

We have obtained an answer which suggests that an area has a negative value. In order to get some idea as to what this might mean let us consider the graphical representation of the area (Fig. 7.5). Our definite integral represents the area

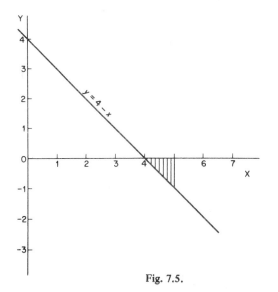

Fig. 7.5.

contained by the straight line $y = 4 - x$, the X-axis and the ordinates $x = 4$ and $x = 5$. From the sketch we see that the area in question lies below the X-axis. It follows then that the solution of the definite integral will be negative if y takes negative values between the limits.

If y takes both positive and negative values between the limits, the definite integral is the algebraic sum of the positive and negative contributions. For instance consider the definite integral

$$\int_2^6 (4 - x)\, dx.$$

This is the same integral as before but the limits have been changed. The integration gives

$$\int_2^6 (4 - x)\, dx = \left[4x - \frac{x^2}{2} \right]_2^6 = (24 - 18) - (8 - 2)$$

$$= 0.$$

The area under the curve, as represented by the definite integral, is shown in Fig. 7.6.

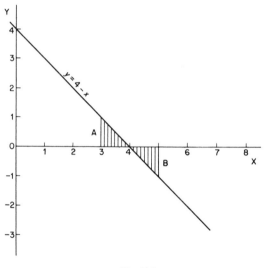

Fig. 7.6.

It is seen from the geometry of the area that the contribution above the X-axis is equal to the contribution below the X-axis and so the algebraic sum of the positive and negative components is zero.

To summarise then, the definite integral may give a positive, zero or negative value according to the values of the limits and the sign of y between those limits.

Sometimes we might wish to know the value of the integral and consider all the areas to be positive. For example in Fig. 7.6 we want to know the value of the area A plus the value of the area B considered positive. In such a case we must perform two integrations, one of which determines the area which lies above the X-axis, the other integral determines the area which lies below the X-axis. In such cases where we consider all areas to be positive we say that area A + area B represent the *whole area* enclosed by the curve of the function, the X-axis and the two ordinates.

Suppose now we wish to integrate between two negative limits, for instance

$$\int_{-4}^{-2} (4 - x)\, dx.$$

In this case values of $(4 - x)$ are positive between the limits of integration.

$$\int_{-4}^{-2} (4 - x)\, dx = \left[4x - \frac{x^2}{2} \right]_{-4}^{-2}$$

$$= (-8 - 2) - (-16 - 8)$$

$$= 14.$$

Even though this area lies to the left of the Y-axis, the area is positive. However, it must be noted that such a positive area will result only if the integration is carried out from left to right, i.e. the lower limit on the X-axis must be to the left of the upper limit.

This last remark introduces us to the general property of the definite integral that reversal of the limits of integration changes the sign of the definite integral. By using the notation of the previous section we have

$$\int_a^b f(x)\, dx = F(b) - F(a).$$

Likewise we may write

$$\int_b^a f(x)\, dx = F(a) - F(b)$$

$$= -\int_a^b f(x)\, dx. \qquad (7.31)$$

Exercises 7.12

Evaluate the following definite integrals:

(1) $\displaystyle\int_{-\frac{\pi}{2}}^{\frac{\pi}{2}} \sin x\, dx$

(2) $\displaystyle\int_{0-}^{10} (x^2 - x + 1)\, dx$

(3) $\displaystyle\int_{-1}^{+1} (1 - x^2)^2\, dx$

(4) $\displaystyle\int_{-2}^{0-} (x - 1)^{-3}\, dx$

(5) $\displaystyle\int_{-1}^{2} (x+2)^{-1}\, dx$ (6) $\displaystyle\int_{-1}^{0} (3x^2 - 2x - 2)\, dx$

(7) $\displaystyle\int_{0}^{1} (x-1)(x+3)\, dx$ (8) $\displaystyle\int (4-x)^{-3}\, dx.$

7.13   RATES OF GROWTH

Suppose that we are doing an experiment in which we are studying the size of a population. The nature of the population need not be specified at this stage; it might be a bacterial culture, the human population in a certain area, the number of plants in a particular region, or it might even refer to the number of leaves on a tree. At a particular point in time (t) let us denote the size of the population by $W(t)$. $W(t)$ is a function of t because its size is obviously changing with time. We know from chapter 5 that differentiation of $W(t)$ with respect to t gives us a formula for the rate of change of the population, i.e.

$$\frac{dW(t)}{dt} = w(t). \tag{7.32}$$

Here $w(t)$ is a function of t and represents the rate of change of the population at any time t. Thus, if we are given the expression $W(t)$ we can calculate $w(t)$ simply by differentiating with respect to t. Suppose now that we are faced with the reverse problem, we are given the expression $w(t)$ for the rate of change of the population and we require to know the value of $W(t)$ the actual size of the population at any time t. First of all we can take expression (7.32) and, according to the definition of an integral, we may write

$$W(t) = \int w(t)\, dt = f(t) + C. \tag{7.33}$$

Here we are faced with the difficulty that we do not know the correct value of the constant C. In order to obtain this we have to know the value of the size of the population at one particular point in time. Suppose that at a particular time T we know the value of $W(t)$ which we denote $W(T)$. We can use this in expression (7.33) to determine C:

$$W(T) = f(T) + C,$$

to give

$$C = W(T) - f(T) \tag{7.34}$$

We now wish to obtain the value of $W(t')$ the size of the population at any other point in time t'. Again, by inserting the value for t' into expression (7.33) we obtain

$$W(t') = f(t') + C,$$

but, since C has now a particular value given by expression (7.34), the above expression becomes

$$W(t') = f(t') + W(T) - f(T). \qquad (7.35)$$

Expression 7.35 may be arranged to take the form

$$W(t') = f(t') - f(T) + W(T).$$

If we now consider expression (7.33) again and use the definition of a definite integral we may write

$$\int_T^{t'} w(t)\, dt = f(t') - f(T) \qquad (7.36)$$

and therefore the size of the population at time t' is given by

$$W(t') = W(T) + \int_T^{t'} w(t)\, dt. \qquad (7.37)$$

Thus in order to obtain a value of the size of the population at a particular time t' we need to know the size of the population at some other point in time and add to it the amount given by performing the integration of the function denoting the rate of change between the limits T and t'. Let us now illustrate what has been said so far by way of an example:

Example I

A population of bacteria is growing at the rate of

$$10^2 \exp[0.1t]$$

individuals per hour. At the start of the experiment there were 10^3 individuals in the population. What was the size of the population one hour from the start of the experiment?

At the start of the experiment let us put $t = 0$. We are told that $W(0) = 10^3$. We want to determine the value of $W(1)$ the size of the population one hour from the start of the experiment. By inserting the appropriate values into (7.37) we write

$$W(1) = \int_0^1 10^2 \exp[0.1\ t]\ dt + 10^3$$

$$= \frac{10^2}{0.1} \exp[0.1\ t] + 10^3$$

$$= 10^3 [\exp[0.1] - 1] + 10^3$$

$$= 10^3 [1.105 - 1.000] + 10^3$$

$$= 1105.$$

(The value of $\exp[0.1]$ to four significant figures was determined from the tables.)

Exercises 7.13

(1) Use the data given in Example I above to determine the increase in the size of the population during the fifth hour after the start of the experiment.

(2) A substance starts to enter the circulation of a cat at a time $t = 0$. According to a theoretical model of the system the rate of increase of the substance within the circulation is given by the formula

$$y' = A \exp[-\theta t] - B \exp[-\lambda t]$$

where y' is the rate of increase of the substance within the circulation at time t and A, B, θ and λ are constants. Determine an expression for the total amount of the substance in the circulation at any time T and express your answer in terms of the constants.

7.14   INTEGRATION BY SUBSTITUTION

In order to evaluate certain integrals of the form $\int f(x)\ dx$ it may be necessary to introduce a substitution by putting x equal to some function of z where z is a new variable. Integration is then performed with respect to z to obtain a function of z which may then be converted back to a function of x, the original variable.

Let

$$I = \int f(x)\ dx. \tag{7.38}$$

We introduce the substitution

$$x = \theta(z)$$

so that

$$f(x) = f[\theta(z)] = F(z).$$

We may differentiate expression (7.38) and obtain

$$\frac{dI}{dx} = f(x) = F(z).$$

By the use of the function of a function theorem we may write

$$\frac{dI}{dz} = \frac{dI}{dx} \cdot \frac{dx}{dz} = F(z).\theta'(z),$$

and therefore

$$I = \int F(z) . \theta'(z) \, dz. \tag{7.39}$$

However, since

$$\theta'(z) = \frac{dx}{dz}$$

we may write

$$I = \int F(z) \frac{dx}{dz} \, dz. \tag{7.40}$$

The usefulness of the above transformation will be illustrated by way of an example:

Example I

Evaluate $\int x \exp[x^2] \, dx.$

This integral is encountered in statistical problems. We have so far not dealt with the integration of a function containing $\exp[x^2]$ although we could easily integrate $\exp[x]$. Let us then make the substitution.

$$z = x^2 \text{ or } x = z^{1/2}$$

so that

$$f(x) = x \exp[x^2]$$

and

$$F(z) = (z)^{1/2} \exp[z].$$

Now

$$\frac{dx}{dz} = \tfrac{1}{2}(z)^{-1/2},$$

so we may write

$$\int f(x)\,dx = \int F(z)\frac{dx}{dz}\,dz$$

$$= \int (z)^{1/2} \exp[z]\tfrac{1}{2}(z)^{-1/2}\,dz$$

$$= \int \tfrac{1}{2}\exp[z]\,dz$$

$$= \tfrac{1}{2}\exp[z] + C.$$

We have previously made the substitution

$$z = x^2$$

and the evaluation of our integral therefore becomes

$$\int f(x)\,dx = \int x \exp[x^2]\,dx$$
$$= \tfrac{1}{2}\exp[x^2] + C.$$

We can also solve definite integrals by this method but we must be careful with the limits. It is often best to determine the indefinite integral and then to insert the limits. Let us consider again the above example but this time as a definite integral.

Example II

Evaluate $\int_0^2 x \exp[x^2]\,dx$.

First of all we may use the result of the first example and write directly

$$\int x \exp[x^2]\,dx = [\tfrac{1}{2}\exp[x^2]]_0^2$$
$$= \tfrac{1}{2}[e^4 - 1]$$

Alternatively we can perform the integration by making the substitution

$$z = x^2$$

as we did previously but we also consider the values of the limits in terms of the variable z. Thus

$$\text{when } x = 2, z = 4,$$

and

$$\text{when } x = 0, z = 0,$$

Therefore

$$\int_0^2 x \exp[x^2] dx = \int_0^4 F(z) \frac{dx}{dz} dz$$

$$= \tfrac{1}{2} \int_0^4 \exp[z] dz$$

$$= \tfrac{1}{2}[e^4 - 1].$$

If we can convert the limits in this way, the process of integration is easier since we do not need to convert back to the original variable.

Example III

Evaluate $\displaystyle\int \frac{x^5}{a^6 + x^6} dx$

In making our choice of substitution, we note that the numerator is $\tfrac{1}{6}$ of the differential coefficient of the denominator. We therefore put

$$z = [a^6 + x^6]$$

from which we obtain

$$\frac{dz}{dx} = 6x^5.$$

In this example we have

$$f(x) = \frac{x^5}{a^6 + x^6}$$

but, from the substitution that we have chosen, it is not immediately obvious what our corresponding function $F(z)$ should be. We could put

$$x = (z - a^6)^{1/6}$$

and obtain

$$F(z) = \frac{(z - a^6)^{5/6}}{z} \qquad (7.41)$$

$$\frac{dx}{dz} = \tfrac{1}{6}(z-a^6)^{-5/6} \text{ and } F(z)\cdot\frac{dx}{dz} = \frac{1}{6z}.$$

According to expression (7.40) our integral will take the form:

$$I = \int F(z)\frac{dx}{dz}\,dz$$

$$= \int \frac{1}{6z}\,dz$$

$$= \tfrac{1}{6}\ln(a^6 - x^6) + C.$$

Although we have gone through the process of integration according to the rules that we have just learned there is a neater way of doing the evaluation which lends itself to this and certain other problems. Instead of using expression (7.41) for F(z) we could have put instead

$$F(z) = \frac{1}{6z}\frac{dz}{dx},$$

and our integral then becomes

$$I = \int \frac{1}{6z}\frac{dz}{dx}\frac{dx}{dz}\,dz$$

$$= \int \frac{1}{6z}\,dz.$$

Example IV

Evaluate $I = \int \sin x \cos x\, dx$.

For this example put

$$z = \cos x$$

so that

$$\frac{dz}{dx} = -\sin x$$

and

$$F(z) = -z\,\frac{dz}{dx}.$$

Substituting these expressions into equations (7.40) yields

$$I = \int -z \, dz$$
$$= -\frac{z^2}{2} + C.$$
$$= \cos^2 x + C.$$

Exercises 7.14

Evaluate the following integrals:

(1) $\int (x)^{-1/2} \exp[x^{1/2}] \, dx$

(2) $\int \frac{x}{(1 - x^2)^{1/2}} \, dx$

(3) $\int e^x . \ln (1 + e^x) \, dx$

(4) $\int (x + 3)^{1/2} \, dx$

(5) $\int x(1 + x^2)^{1/2} \, dx$

(6) $\int \frac{e^x}{1 + e^x} \, dx$

(7) $\int \sin^3 x \cos x \, dx$

(8) $\int \tan x \, dx$ i.e. $\int \frac{\sin x}{\cos x} \, dx$

(9) $\int \frac{x^3}{x^4 - 1} \, dx$

(10) $\int x(a^2 + x^2)^3 \, dx$

(11) $\int \cos^3 x \sin x \, dx$

(12) $\int \frac{x}{(a^2 - x^2)^3} \, dx$

(13) $\int x^2(a^3 - x^3)^3 \, dx$

(14) $\int \frac{x}{a^4 - x^4} \, dx$

(15) $\int \frac{\ln x}{x} \, dx$

(16) $\int \frac{(1 + \ln x)^3}{x} \, dx$

(17) $\int \frac{\sin x}{(1 - 2 \cos x)^2} \, dx$

(18) $\int \frac{(\ln x)^n}{x} \, dx$

(19) $\int e^x(1 - e^x)^3 \, dx$

(20) $\int \frac{dx}{x(1 + \ln x)^2}$

(21) $\int \frac{x^3 \, dx}{a^8 - x^8}$

(22) $\int x \exp[-x^2] \, dx$

7.15   INTEGRATION BY PARTS

In a previous chapter on differentiation we found that the differential coefficient of a product is expressed as follows

$$\frac{d(uv)}{dx} = u\frac{dv}{dx} + v\frac{du}{dx} \qquad (7.42)$$

where u and v are functions of x.

If we integrate this expression and put the arbitrary constant equal to zero we obtain

$$(uv) = \int u\frac{dv}{dx}\,dx + \int v\frac{du}{dx}\,dx.$$

We can rearrange this equation to take the form

$$\int u\frac{dv}{dx}\,dx = uv - \int v\frac{du}{dx}\,dx \qquad (7.43)$$

This is known as the formula for *integration by parts*. The integral on the right hand side may often be much easier to solve than the left hand side. This formula is often used when the expression to be integrated contains functions such as $(\ln x)$ or an exponential function.

Example I

Evaluate $\int x^3 (\ln x)\,dx$.

We put

$$u = \ln x \text{ and } \frac{dv}{dx} = x^3$$

By differentiating u and integrating dv/dx we obtain

$$\frac{du}{dx} = \frac{1}{x} \text{ and } v = \frac{x^4}{4}$$

According to the formula for integration by parts we write

$$\int x^3 (\ln x)\,dx = \frac{x^4}{4} \ln(x) - \int \frac{x^4}{4} \frac{1}{x}\,dx$$

$$= \frac{x^4}{4} \ln(x) - \int \frac{x^3}{4}\,dx$$

$$= \frac{x^4}{4} \ln(x) - \frac{x^4}{16} + C.$$

Example II

Evaluate $\int x \cdot e^x\,dx$

Put

$$u = x \text{ and } \frac{dv}{dx} = e^x$$

so that

$$\frac{du}{dx} = 1 \text{ and } v = e^x.$$

By the use of the formula we may write

$$\int x \cdot e^x\,dx = x \cdot e^x - \int e^x\,dx$$

$$= x \cdot e^x - e^x + C.$$

Suppose that we had chosen to put

$$u = e^x \text{ and } \frac{dv}{dx} = x,$$

then

$$\frac{du}{dx} = e^x \text{ and } v = \tfrac{1}{2} x^2$$

and our integral would become

$$\int x \cdot e^x\,dx = \tfrac{1}{2} x^2 \cdot e^x - \tfrac{1}{2} \int x^2 \cdot e^x\,dx.$$

In this case we have not succeeded in simplifying the integral. This is an illustration of the point that if the integral is soluble by the method of integration by parts, we may initially choose the wrong functions for u and v and several attempts may be necessary.

Example III

Evaluate $\int x^2 e^x \, dx$

In this case we put

$$u = x^2 \text{ and } \frac{dv}{dx} = e^x$$

so that

$$\frac{du}{dx} = 2x \text{ and } v = e^x$$

We can now write

$$\int x^2 . e^x \, dx = x^2 . e^2 - 2 \int x . e^x \, dx.$$

The integral on the right hand side, although it is simpler than the one on the left hand side, cannot be written down at once. It is in fact the integral which we evaluated in the previous example so that we may immediately write

$$\int x^2 . e^x \, dx = x^2 . e^x - 2[x . e^x - e^x] + C.$$
$$= e^x[x^2 - 2x + 2] + C.$$

This is an example of integrating by the method of *successive reduction*. Many integrals can only be evaluated by stages which successively produce an integral which is simpler than that of the previous stage. The final stage produces an integral that can be directly evaluated.

Example IV

Evaluate $\int \ln x \, dx$

This example illustrates a useful trick that can be employed in the integration where, at first sight, it would appear that the method of integration by parts is not possible. We put

$$u = \ln x \text{ and } \frac{dv}{dx} = 1$$

so that

$$\frac{du}{dx} = \frac{1}{x} \text{ and } v = x.$$

Our integral then takes the form

$$\int \ln x \, dx = x \cdot \ln x - \int dx$$
$$= x \cdot \ln x - x + C.$$

Exercises 7.15

Integrate the following functions

(1) $x^4 \ln (x)$

(2) $x^n \ln (x)$

(3) $x \sin (ax)$

(4) $x \cos x$

(5) $x^3 \cdot e^x$

(6) $x \exp[ax]$

(7) $x^2 \cdot \exp[-ax]$

(8) $x^2 \sin x.$

Evaluate the following definite integrals:

(9) $\int_0^6 x^2 \ln (2x) \, dx$

(10) $\int_0^{\pi/2a} x \sin (ax) \, dx$

(11) $\int_0^1 x^3 \cdot e^x \, dx$

(12) $\int_0^1 x \cdot \exp[-x] \, dx$

(13) Baranov developed expressions for commercial yields of fish in terms of the lengths of the fish. His formula for the weight of the commercial population is

$$W = \int_{L'}^{\infty} NbL^3 \exp[-iL] \, dL$$

where N, b and i' are constants. Show that

$$W = \frac{bL^3 N \exp[-iL]}{i'} \left\{ 1 + \frac{3}{i'L} + \frac{6}{(i'L)^2} + \frac{6}{(i'L)^3} \right\}.$$

7.16 INTEGRATION OF PARTIAL FRACTIONS

Suppose we wish to integrate a function such as

$$\frac{2}{x^2 - 1}.$$

At first sight this may not be done by making use of any of the methods so far discussed. Closer inspection of the function shows that it may be rearranged as follows:

$$\frac{2}{x^2 - 1} = \frac{2}{(x - 1)(x + 1)} = \frac{1}{(x - 1)} - \frac{1}{(x + 1)}$$

The integration of the function in the form of two separate functions is now easily carried out. We have taken our original function and resolved it into *partial fractions* prior to integration. In order to resolve a function into partial fractions, the denominator must be such that it can be expressed as the product of factors of the form $(ax + b)$. Also if the numerator is of higher degree or the same degree as the denominator, preliminary division must be carried out and the remainder expressed as partial fractions.

Example I

Evaluate $\displaystyle\int \frac{dx}{2 - x - x^2}$

We can factorise the denominator of the function to be integrated and write

$$\frac{1}{2 - x - x^2} = \frac{1}{(x + 2)(1 - x)}.$$

We now choose two constants A and B so that the function may be separated as follows:

$$\frac{1}{(x + 2)(1 - x)} = \frac{A}{(x + 2)} + \frac{B}{(1 - x)}$$

In order to determine the values of A and B we multiply throughout the above equation by $(x + 2)(1 - x)$ and obtain

$$1 = A(1 - x) + B(x + 2).$$

This expression holds for all values of x so we put $x = 1$ to determine B, and put $x = -2$ to determine A. Thus

$$B = \tfrac{1}{3} \text{ and } A = \tfrac{1}{3}.$$

We can now solve the integral since it is put in the form

$$\int \frac{dx}{2 - x - x^2} = \int \frac{dx}{3(x + 2)} + \int \frac{dx}{3(1 - x)}$$

$$= -\tfrac{1}{3} \ln(1 - x) + \tfrac{1}{3} \ln(x + 2) + C.$$

$$= \tfrac{1}{3} \ln \frac{(x + 2)}{(1 - x)} + C.$$

Example II

Evaluate $\displaystyle\int \frac{x^3}{1 - x^2}\, dx$

In this case the numerator is of higher degree than the denominator and so division must be performed first of all.

$$\frac{x^3}{x^2 - 1} = x + \frac{x}{x^2 - 1}$$

Next we choose constants A and B so that

$$\frac{x}{x^2 - 1} = \frac{A}{x - 1} + \frac{B}{x + 1}$$

or, after cross multiplication,

$$x = A(x + 1) + B(x - 1).$$

If we put $x = -1$ we obtain $B = \tfrac{1}{2}$, and if we put $x = 1$ we obtain $A = \tfrac{1}{2}$ so that

$$\frac{x^3}{x^2 - 1} = x + \frac{1}{2(x - 1)} + \frac{1}{2(x + 1)}$$

We are now able to solve the integration:

$$\int \frac{x^3}{1 - x^2}\, dx = -\frac{x^2}{2} - \tfrac{1}{2} \ln (x - 1)(x + 1) + C.$$

Example III

Evaluate $\displaystyle \int \frac{x + 1}{(x - 1)^2}\, dx$.

The integrand is resolved into partial fractions in the following way:

$$\frac{x + 1}{(x - 1)^2} = \frac{A}{(x - 1)} + \frac{B}{(x - 1)^2}$$

These are the only two functions whose denominators could have an l.c.m. = $(x - 1)^2$. We cross multiply and obtain

$$x + 1 = A(x - 1) + B.$$

We put $x = 1$ to obtain $B = 2$, and put $x = 0$ to obtain $A = 1$. Our integral now takes the form

$$\int \frac{x + 1}{(x - 1)^2}\, dx = \int \frac{dx}{(x - 1)} + 2 \int \frac{dx}{(x - 1)^2}$$

$$= \ln (x - 1) - \frac{2}{(x - 1)} + C.$$

Exercises 7.16

Integrate the following:

(1) $\displaystyle \frac{6x}{(x + 1)(x - 2)}$

(2) $\displaystyle \frac{x^2 + 2}{x^2 - 1}$

(3) $\displaystyle \frac{2x + 1}{x^3 + 2x^2 - 3x}$

(4) $\displaystyle \frac{1}{x^2 + x - 6}$

(5) $\displaystyle \frac{1}{x^2 + x - 20}$

(6) $\displaystyle \frac{x}{x^2 - 1}$

(7) $\dfrac{x^2}{x^2-4}$

(8) $\dfrac{1}{x^2-(a+b)x+ab}$

(9) $\dfrac{2x-5}{x^2-5x+6}$

(10) $\dfrac{5x+2}{x^2-4x+4}$

(11) $\dfrac{1}{3x^2-10x+3}$

(12) $\dfrac{1}{x^2-7^2}$

(13) $\dfrac{3}{x^2-x-30}$

(14) $\dfrac{1}{3x^2-x-2}$

(15) $\dfrac{x^3}{x^2+x-20}$

(16) $\dfrac{1}{x(5-2x)}$

7.17   THE PROBABILITY INTEGRAL

If you have worked through the previous sections of this chapter you might think that you are now competent to integrate most of the simple functions that might arise. However there are certain integrals which cannot be evaluated by any simple methods that have so far been described. Such integrals may take the form

$$\int \ln(\sin x)\,dx,\ \int (1+x^3)^{1/2}\,dx \text{ or } \int \exp[-x^2]\,dx.$$

The last of these integrals is very important in statistical work. It is known as the *probability integral*. Its use in statistics is so widespread that nearly every text-book on the subject contains tables of values of the definite integral

$$P = \frac{1}{\sqrt{2\pi}} \int_0^t \exp\left[-\frac{t^2}{2}\right]dt \qquad (7.44)$$

for various values of t. The integral is tabulated in this form because it arises in this way in statistical problems. Sometimes we need to evaluate an integral of the form

$$I = \int x^n \exp[-\tfrac{1}{2}x^2]\,dx \qquad (7.45)$$

where n is a positive integer. In order to solve the integral we must reduce it to the form of expression (7.44) and look up the appropriate definite integral in the tables, or look to see if the integral can be solved outright as for the case of the similar Example I of Section 7.14.

Let us consider the various possible values of n:

(a) $n = 1$

$$I = \int x \exp[-\tfrac{1}{2}x^2]\, dx.$$

According to the methods of Section 7.14 we put

$$z = \tfrac{1}{2} x^2 \text{ so that } \frac{dz}{dx} = x,$$

and

$$F(z) = \frac{dz}{dx} \exp[-z].$$

We insert these expressions into (7.40) and obtain

$$I = \int \exp[-z]\, dz$$
$$= -\exp[-z] + C$$
$$= -\exp[-\tfrac{1}{2}x^2] + C. \qquad (7.46)$$

(b) $n = 3$

$$I = \int x^3 \exp[-\tfrac{1}{2}x^2]\, dx.$$

We start off by making the same substitution as we did for case (a) above

$$z = \tfrac{1}{2} x^2 \text{ and } \frac{dz}{dx} = x.$$

For this case we have

$$F(z) = (2z)\frac{dz}{dx} \exp[-z].$$

We insert these expressions into 7.40 and obtain

$$I = 2\int z \exp[-z]\, dz. \qquad (7.47)$$

The integral is now solved by the method of integration by parts.
We put

$$u = z \text{ and } \frac{dv}{dz} = \exp[-z]$$

so that

$$\frac{du}{dz} = 1 \text{ and } v = -\exp[-z].$$

Our integral now takes the form

$$I = -2z \exp[-z] + 2 \int \exp[-z] \, dz$$
$$= -2z \exp[-z] - 2 \exp[-z]$$
$$= -2(1 + \tfrac{1}{2}x^2) \exp[-\tfrac{1}{2}x^2]. \qquad (7.48)$$

(c) $n = 2m + 1$ (m is a positive integer greater than 1).

$$I = \int x^{(2m+1)} \exp[-\tfrac{1}{2}x^2] \, dx.$$

We proceed as before and put

$$z = \frac{x^2}{2} \text{ so that } \frac{dz}{dx} = x.$$

but in this case

$$F(z) = 2^m . z^m . \exp[-z] . \frac{dz}{dx}$$

and our integral reduces to

$$I = 2^m \int z^m \exp[-z] \, dz.$$

We now try the method of integration by parts and put

$$u = z^m \text{ and } \frac{dv}{dz} = \exp[-z]$$

so that

$$\frac{du}{dz} = mz^{(m-1)} \text{ and } v = -\exp[-z],$$

which, upon substitution into expression (7.43), yields

$$I = 2^m \left\{ -z^m \exp[-z] + \int mz^{(m-1)} \exp[-z] \, dz \right\}. \qquad (7.49)$$

This integral is now only soluble if $m = 2$ because in this particular case the integral on the right hand side is identical with the integral of (7.47) which we have already evaluated. If m exceeds 2 we must take the integral of the right hand side of (7.49) and by the method of integration by parts, reduce the power of z in the integrand to $(m - 2)$. This process is repeated, so that successive reductions in the power of z eventually reduce the integral to the form of expression (7.47). Thus for odd values of n, the integral of (7.45) can be evaluated.

(d) $n = 2m$ (m is a positive integer)

$$I = \int x^{2m} \exp\left[-\tfrac{1}{2}x^2\right] dx.$$

Let us try the method of integration by parts and put

$$u = \exp\left[-\tfrac{1}{2}x^2\right] \text{ and } \frac{dv}{dx} = x^{2m}$$

so that

$$\frac{du}{dx} = -x\exp\left[-\tfrac{1}{2}x^2\right] \text{ and } v = \frac{x^{(2m+1)}}{(2m+1)}$$

Our integral now takes the form

$$\int x^{2m} \exp\left[-\tfrac{1}{2}x^2\right] dx = \frac{x^{(2m+1)}}{(2m+1)} \exp\left[-\tfrac{1}{2}x^2\right]$$
$$+ \frac{1}{(2m+1)} \int x^{(2m+2)} \exp\left[-\tfrac{1}{2}x^2\right] dx.$$

At first sight it would appear that we have not simplified our integral but instead have raised the power of x in the integrand by 2. Let us rearrange the above expression so that it takes the form

$$\int x^{(2m+2)} \exp\left[-\tfrac{1}{2}x^2\right] dx = (2m+1)\int x^{2m} \exp\left[-\tfrac{1}{2}x^2\right] dx$$
$$- x^{(2m+1)} \exp\left[-\tfrac{1}{2}x^2\right]. \qquad (7.50)$$

In this expression we have in effect reduced the power of x in the left hand integral. If we start off with the integral on the left hand side we see that it is equal to the expression on the right hand side which contains an integral in which the power of x has been reduced by two.

If now, in expression (7.50), we replace m by $(m-1)$ we obtain

$$\int x^{(2m)} \exp\left[-\tfrac{1}{2}x^2\right] dx = (2m-1)\int x^{(2m-2)} \exp\left[-\tfrac{1}{2}x^2\right] dx$$
$$- x^{(2m-1)} \exp\left[-\tfrac{1}{2}x^2\right] \qquad (7.51)$$

The left hand side is our original integral and it has been transformed into an integral in which the power of x has been reduced by 2. The integral on the right is evaluated by using expression (7.51) but replacing m by $(m-1)$. This is repeated until an expression is finally obtained in which the integral is of the form

$$\int \exp[-\tfrac{1}{2}x^2]\,dx, \tag{7.52}$$

and this of course can only be evaluated by looking up the tables.

The cases where n is negative or non-integer cannot usually be evaluated or converted to the form (7.52) by methods that are within the scope of this book.

Exercises 7.17

(1) Evaluate $\int_0^\infty x\,.\,\exp[-x^2]\,dx$ (2) Integrate $\cos x \ln(\sin x)$

(3) Evaluate $\int_1^3 x\,\exp[-\tfrac{1}{2}x^2]\,dx$.

7.18 NUMERICAL INTEGRATION

Sometimes we may need to calculate the area under a curve and the usual methods of integration may not be available to us. Such a situation may arise in two ways:

(1) The curve may have an equation such as

$$y = x^{1/2}\exp[-x^2].$$

As we have seen from the previous section, such a function cannot be integrated readily. In this case we have to calculate the values of y for various values of x within the limits of the integration, draw the curve and calculate the area by dividing it up into smaller parts whose shapes are reasonable approximations to simpler geometric shapes.

(2) Cases in which we may have plotted a set of experimental points which have not been fitted to any formula, and we want to determine the area under some portion of the curve.

Two methods of determining the area under a curve are now considered.

(1) *The Trapezoidal Rule*

We consider the definite integral

$$A = \int_a^b f(x)\,dx. \tag{7.53}$$

In all cases of numerical integration it is best to make a sketch of the curve as for example in Fig. 7.7.

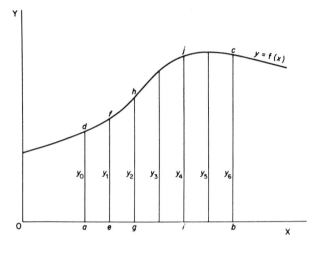

Fig. 7.7.

The required area, as determined by the limits of integration is bounded by the two ordinates ad and bc, the part ab along the X-axis and the curved part cd. The area is divided into vertical strips by erecting ordinates equally spaced along ab. The resultant n strips are thus of equal width. The ordinates are labelled $y_0, y_1, y_2, \ldots y_n$, and in the case of Fig. 7.7 we have chosen $n = 6$. Now consider the area of the first strip $aefd$. If the strip is sufficiently narrow, the portion of the curve fd which bounds it is virtually a straight line so that the area $aefd$ approximates to a trapezium and its area δA_1 is given by

$$\delta A_1 = \tfrac{1}{2} h(y_0 + y_1) \tag{7.54}$$

where h is the width of the strip.

In a similar way we can calculate the areas of the other strips in terms of the lengths of their bounding ordinates and the strip width. The total area is the sum of the individual strip areas:

$$A = \tfrac{1}{2} h(y_0 + y_1) + \tfrac{1}{2} h(y_1 + y_2) \ldots$$
$$+ \tfrac{1}{2} h(y_{n-1} + y_n)$$
$$= \tfrac{1}{2} h(y_0 + y_n) + h(y_1 + y_2 + \ldots y_{n-1}),$$

or in words, we may write:

Area = width of strip $[\tfrac{1}{2}$(first ordinate + last ordinate)
                      + sum of intervening ordinates]. $\tag{7.55}$

The accuracy of the result increases with the number of strips that is chosen. In every example the number of strips must be such that each strip does in fact approximate closely to a trapezium.

(2) Simpson's Rule

Consider again the area of Fig. 7.7. We have divided it into six strips of equal width. Let us suppose that the strips are not really very good approximations to trapeziums. We could either divide the area into a greater number of strips, and this would make the calculation more tedious, or approximate each strip to a different, but more accurate, shape. By the method of Simpson's Rule, each arc of the curve which bounds the top of a strip is considered to approximate to a line whose equation is

$$y = Ax^2 + Bx + C,$$

where A, B, and C are constants for the particular arc that is being considered. Let us not worry about the magnitude of these constants but consider the combined area of the two adjacent strips which form the area $aghd$ in the figure. In terms of the constants the area of these strips is found by integration as follows:

$$\text{Area } aghd = \int_a^{a+2h} (Ax^2 + Bx + C)\,dx$$

$$= \left[\frac{Ax^3}{3} + \frac{Bx^2}{2} + Cx\right]_a^{a+2h}$$

$$= \frac{A}{3}(6a^2h + 12ah^2 + 8h^3)$$

$$+ \frac{B}{2}(4ah + 4h^2) + 2Ch. \tag{7.56}$$

We can also obtain values of the ordinates y_0, y_1 and y_2, which bound our two strips, in terms of the constants A, B, and C.

$$y_0 = Aa^2 + Ba + C,$$
$$y_1 = A(a + h)^2 + B(a + h) + C,$$

and

$$y_2 = A(a + 2h)^2 + B(a + 2h) + C.$$

We may combine these values in the following way:

$$\tfrac{1}{3}(y_0 + 4y_1 + y_2) = \frac{A}{3}(6a^2 + 12ah + 8h^2)$$

$$= \frac{B}{2}(4a + 4h) + 2C.$$

If we compare this expression with (7.56) above we see that

$$\text{Area } aghd = \frac{h}{3}(y_0 + 4y_1 + y_2).$$

We can repeat this process to get a value for the area of the next pair of strips which form the area *gijh*:

$$\text{Area } gijh = \frac{h}{3}(y_2 + 4y_3 + y_4).$$

Likewise the last pair of strips have an area given by

$$\text{Area } bcji = \frac{h}{3}(y_4 + 4y_5 + y_6).$$

The total area for the whole figure is now obtained:

$$A = \frac{h}{3}(y_0 + 4y_1 + 2y_2 + 4y_3 + 2y_4 + 4y_5 + y_6).$$

In order to calculate an area by Simpson's Rule the area must be divided up into an even number of strips. A general formulation of the area by the method is

$$\text{Area} = \tfrac{1}{3}(\text{strip width}) \; [(\text{first ordinate} + \text{last ordinate})$$
$$+ 2(\text{sum of intervening even ordinates})$$
$$+ 4(\text{sum of odd ordinates})] . \qquad (7.57)$$

Example

Calculate the value of $\int_0^{0.6} x^3 \, dx$

by the Trapezoidal and Simpson's Rules taking $h = 0.1$. First of all we draw up a table of values of x^3 for corresponding values of x within the limits of the integration:

x	0	0.1	0.2	0.3	0.4	0.5	0.6
x^3	0	0.001	0.008	0.027	0.064	0.125	0.216

We now work out the following values:

Sum of the first and last ordinates $= 0.216$
Sum of odd ordinates $= 0.153$
Sum of intervening even ordinates $= 0.072$

(a) by the Trapezoidal Rule

Area = width of strip $[\frac{1}{2}$(first + last ordinates)
$+$ sum of intervening ordinates$]$
$= 0.1[0.108 + 0.225] = 0.0333.$

(b) by Simpson's Rule

Area $= \frac{1}{3}$(width of strip) $[$(first + last ordinates)
$+ 2$(sum of intervening even ordinates
$+ 4$(sum of odd ordinates)$]$

$= \frac{1}{3}(0.1)[0.216 + 0.144 + 0.612]$
$= \frac{1}{3}(0.1)(0.972)$
$= 0.0324.$

We can check our answers by integration

$$\int_0^{0.6} x^3 \, dx = \left[\frac{x^4}{4}\right]_0^{0.6} = \frac{0.1296}{4} = 0.0324.$$

As expected the method of Simpson's rule gives the more accurate answer, but by the trapezoidal rule the inaccuracy is only about 3% for this particular example.

Exercises 7.18

(1) Use the Trapezoidal rule and Simpson's rule to evaluate

$$\int_1^2 (2x^3 + 3x^2 + 5) \, dx$$

and compare your answers with the exact answer found by direct integration.

(2) A set of experimental results is given in the following table:

x	0	1	2	3	4	5	6	7	8	9	10
y		10.1	15.2	16.3	16.5	16.4	16.3	16.0	15.8	15.5	15.2	12.1

Use Simpson's Rule to evaluate $\int_0^{10} y\, dx$

(3) Evaluate by Simpson's Rule $\int_0^1 \exp[-x]\, dx$

(divide the area into 10 strips).

(4) Evaluate the following integrals by Simpson's Rule

$$\int_1^{1.5} \frac{dx}{x},\ \int_1^2 \frac{dx}{x},\ \int_1^3 \frac{dx}{x},$$

and compare your answers with the values of ln 1.5, ln 2 and ln 3 found in the tables at the end of this book.

8

DIFFERENTIAL EQUATIONS

8.1 DEFINITIONS

If we are given the equation

$$\frac{dy}{dx} = 3x^2 \tag{8.1}$$

and asked to find y, we could obtain the answer simply by finding the indefinite integral

$$y = \int 3x^2 \, dx$$

and obtaining the answer

$$y = x^3 + C.$$

The equation (8.1) which is a relationship involving dy/dx is an example of a *differential equation*. Any differential equation of the form

$$\frac{dy}{dx} = f(x)$$

is already familiar to us because the solution is obtained simply by integrating $f(x)$ to obtain

$$y = \int f(x) \, dx = F(x) + C \tag{8.2}$$

and the equation (8.1) is an example of this form of differential equation.

Many differential equations do not give a simple solution. Some of them can be so difficult that their solution is best left to the professional mathematician. Fortunately, many of the differential equations which are relatively simple to solve are useful to the biologist and we will consider some of these in this chapter.

One point worth remembering is that, as in the case of the evaluation of indefinite integrals, the correctness of our answer can always be verified by differentiation and substitution back into the original equation.

Differential equations which involve dy/dx and one or both of the quantities x and y are known as differential equations of the first order and their solution will involve one arbitrary constant as in the case of the solution (8.2) above.

Differential equations which involve d^2y/dx^2 and a combination of the quantities dy/dx, x and y are known as differential equations of the second order. An example of such an equation is

$$\frac{d^2 y}{dx^2} + 6x \frac{dy}{dx} + (x - y)^2 = 0$$

and the solution of this equation, and all second order differential equations will involve two arbitrary constants.

One can also define third, fourth . . . and nth order differential equations whose corresponding solutions will involve three, four . . . and n arbitrary constants.

There is no general method for the solution of differential equations. As for the case of integration particular cases must be chosen which illustrate various methods that may be employed.

The study of differential equations can be taken to very high levels. In a book of this kind only a few types can be considered but the reader is encouraged to refer to more specialist texts if he wishes to continue beyond the level of this book.

8.2   DIFFERENTIAL EQUATIONS OF THE FORM

$$\frac{dy}{dx} = f(y)$$

In order to solve this equation we remember the result of Section 5.13 viz

$$\frac{dx}{dy} = \frac{1}{dy/dx}$$

and apply it to the present case and obtain

$$\frac{dx}{dy} = \frac{1}{dy/dx} = \frac{1}{f(y)}.$$

From the definition of the indefinite integral we may write

$$x = \int \frac{dy}{f(y)} \tag{8.3}$$

Example I

Solve the differential equation

$$\frac{dy}{dx} = \frac{1}{y}.$$

We use equation (8.3), put $f(y) = 1/y$, and obtain

$$x = \int y\,dy = \frac{y^2}{2} + C. \qquad (8.4)$$

In this solution we may get rid of the constant C if we are told certain conditions. For instance, if we are given the example:

$$\text{Solve } \frac{dy}{dx} = \frac{1}{y} \text{ given that } y = 1 \text{ when } x = 0.$$

We obtain the solution (8.4) above but now we insert the given values of x and y into the solution and obtain

$$0 = \tfrac{1}{2} + C$$

or

$$C = -\tfrac{1}{2}.$$

The complete solution of the differential equation is, in this particular case,

$$x = \tfrac{1}{2}(y^2 - 1).$$

Example II

Solve the differential equation

$$\frac{dy}{dx} = Ky \ (K = \text{constant}). \qquad (8.5)$$

This is one of the most important and frequently occurring differential equations that the biologist is likely to encounter. If we apply the result of equation (8.3) we obtain for this case

$$x = \int \frac{dy}{Ky} = \frac{1}{K} \ln y + C$$

which, after re-arrangement takes the form

$$\ln y = K(x - C).$$

If we take exponentials of each side we obtain

$$y = \exp[Kx - KC]$$

or

$$y = A \exp[kx]$$

where $A = \exp[-KC]$ is a constant.

This solution, or variations of it, occurs in studies on population growth, radioactive decay, heat loss from a body, elementary reaction kinetics, the diffusion of ions across a membrane etc.

Exercises 8.2

Solve the following differential equations and check your solutions by differentiation:

(1) $\dfrac{dy}{dx} = (y - K)$ K is a constant   (2) $\dfrac{dy}{dx} = \exp[-3y]$

(3) $\dfrac{dy}{dx} = ax + by$ (Hint: obtain a differential equation in terms of z and dz/dx where $z = (ax + by)$).

(4) A cell membrane has a capacitance (C) and resistance (R). The equation connecting the charge (q) with the transmembrane potential (E) is

$$E = R\,\frac{dq}{dt} + \frac{q}{C}$$

If $q = Q$ when $t = 0$, show that for any subsequent value of t

$$q = EC - (EC - Q)\exp[-t/RC]$$

(5) In a certain type of enzyme-controlled reaction the rate of breakdown of the substrate is given by

$$\frac{dx}{dt} = \frac{K_3[(S_0) - x]}{1 + \{[(S_0) - x]/K_s\} + (x/K_t)}$$

where $[S_0 - x]$ is the concentration of substrate remaining after a time t, S_0 is the initial concentration of the substrate, and the other quantities are constants.

Show that

$$K_3 = \left\{ \frac{1}{K_s} - \frac{1}{K_i} \right\} \cdot \frac{x}{t} + [1 + S_0/K_i] \frac{1}{t} \ln \left\{ \frac{S_0}{S_0 - x} \right\}.$$

8.3 RADIOACTIVE DECAY

The rate at which any radioactive substance decomposes at any point in time is proportional to the amount of undecomposed substance present at that point in time.

Let $N = f(t)$ be the amount of undecomposed substance present at time t. The function dN/dt represents the rate of increase of the substance at that time. A radioactive substance, however, *decays* with time and so dN/dt is negative. We set up our differential equation as follows:

$$\frac{dN}{dt} = -kN \qquad (8.6)$$

where k is a positive constant and the negative sign denotes the decrease in N.

Equation 8.6 is really the same as (8.5) in the previous section. We solve this equation in a similar way:

$$\frac{dt}{dN} = -\frac{1}{kN}$$

$$t = -\frac{1}{k} \int \frac{dN}{N} = -\frac{1}{k} \ln N + C.$$

This result is usually expressed in the form

$$N = A \exp[-kt] \qquad (8.7)$$

where A is a constant. If, at $t = 0$, $N = N_0$ our constant A is replaced by N_0 and our solution takes the form

$$N = N_0 \exp[-kt] \qquad (8.8)$$

8.4 HEAT LOSS FROM A HOT BODY

One important problem that may face an animal is the amount of heat that is lost from its body surface if it is exposed to a low temperature. The animal will lose heat at a rate given by Newton's Law of Cooling:

$$\frac{dQ}{dt} = kA.(T_b - T_a) \tag{8.9}$$

where dQ/dt is the rate of heat loss from the body, k is a negative constant depending upon the type of body surface, A is the body surface area, T_b is the body surface temperature and T_a is the ambient temperature.

Many problems involving differential equations are really two separate problems. Firstly there is the problem of setting up the differential equation. This can be very difficult especially if several factors are involved when it is often a case of making simplifications in order to get a manageable equation. Secondly there is the problem of the actual solution of our differential equation. While the latter problem might well be placed in the hands of a professional mathematician, it is usually left to the biologist to formulate the equation. In order to do this it is often best to start from the simplest system even though this might involve making gross assumptions.

Take the case of the animal that is exposed to a low temperature. The animal loses heat to the surroundings by way of the body surface and the respiratory pathways. It gains heat through the metabolic processes within its body and this will vary according to the state of activity of the animal. Further complications arise in that the body surface is not uniform—some parts will lose heat more rapidly than others; the body surface temperature is not constant over the whole of the body; the internal temperature is not constant or uniform; the internal metabolism will not be constant.

In spite of these complications let us take the case of an animal mass M. In addition to its mass, the amount of heat Q contained within the animal is dependent upon its body temperature T_B and specific heat C:

$$Q = MC.(T_B + T_0)$$

where T_0 is a constant such that the heat content of the body at $0°C$ is MCT_0.

We have assumed that if T_B is measured in degrees centigrade then, within the physiological range of temperature, the total heat within the body is a linear function of T_B and

$$T_B = \frac{Q}{MC} - T_0$$

If we make the assumption that $T_B = T_b$, our differential equation takes the form

$$\frac{dQ}{dt} = \frac{kA}{MC}(Q - T')$$

where T' is a constant $= MC(T_a + T_0)$.

The equation is solved by firstly inverting it to take the form

$$\frac{dt}{dQ} = \frac{1}{K(Q - T')}$$

where $K = kA/MC$, and then integrating to obtain

$$Kt = \ln\left[\text{const.}\,(Q - T')\right].$$

We take exponentials of each side so that the solution takes the form

$$\exp[Kt] = [\text{const.}\,(Q - T')]$$

or

$$Q = A\exp[Kt] + T'.$$

This equation gives the amount of heat contained within the animal as a function of time. A value of the constant A could be derived by inserting the appropriate values for the time $t = 0$.

This result is obtained by taking the simplest possible differential equation. In this particular animal heat would be lost to the surroundings until the animal had cooled down to the temperature of the surroundings. In other words we have inserted nothing into the equation which takes into account the heat that is produced within the animal by way of its internal metabolism. Let us suppose that the rate of production of heat in this way is a function of time. Our original differential equation will now have the form

$$\frac{dQ}{dt} = kA.(T_b - T_a) + f(t)$$

and since we may write

$$F(Q) = kA(T_b - T_a),$$

our equation is of the form

$$\frac{dQ}{dt} = F(Q) + f(t) \tag{8.10}$$

Methods of solution of some of the differential equations which are examples of this general form are dealt with in future sections of this chapter.

8.5 PASSAGE OF SUBSTANCES ACROSS A BIOLOGICAL MEMBRANE

One of the most important problems in biology is the study of the passage of substances across thin membranes. Such transport may be entirely passive and be dependent upon purely physical factors or it may be an active process which is dependent upon energy transforming processes within the living cell membrane. Let us consider a thin walled membrane which allows the passage of a solute across it but, in the present case, water does not pass across. Let us represent it in the simple form illustrated in Fig. 8.1.

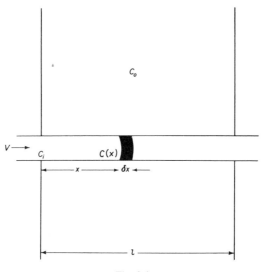

Fig. 8.1.

The membrane forms the wall of a uniform cylindrical vessel of length l and radius a. The concentration of a particular solute at a distance x along the tube from one end is $C(x)$. The concentration of solute in the surrounding medium is C_o and we consider its volume to be so large that the concentration is not affected by any gains or losses to the fluid in the cylinder. The concentration of the fluid as it enters the cylinder is C_i and this is considered to be less than C_o so that the fluid in the cylinder will gain solute from the surrounding fluid by way of osmotic forces. The solution passes along the vessel at a steady rate V.

Let us consider the passage of the solute across the walls of the part of the cylinder indicated in the figure which has a length l. The rate of diffusion across the walls of this part of the cylinder depends upon the area of the wall of the vessel across which the diffusion takes place, and the concentration difference existing across the walls between the fluid inside the cylinder and the fluid in the outside medium.

Thus for our cylinder length δx the amount of substance that passes from the outside to the inside in unit time is given by

$$\delta A = k \, 2\pi a \, . \, \delta x \, . \, (C_0 - C(x))$$

where $2\pi a \delta x$ is the surface area available for diffusion and k is a constant characteristic of the properties of the wall of the vessel and the particular diffusing substance that is being considered. At the same time a volume V of the solution passes through the small cylinder and as it does so its concentration will increase by an amount $\delta C(x)$ where

$$\delta C(x) = \frac{\delta A}{V} = \frac{k 2\pi a}{V} \, \delta x (C_0 - C(x)).$$

In the limit we may write

$$\frac{dC(x)}{dx} = \frac{k 2\pi a}{V} . (C_0 - C(x)).$$

We invert this equation and then integrate to obtain

$$x = \frac{V}{k 2\pi a} \int \frac{dC(x)}{(C_0 - C(x))}$$

$$= - \frac{V}{k 2\pi a} \ln (C(x) - C_0) + C.$$

We put this equation in exponential form and obtain

$$\exp \left[- \frac{k 2\pi a x}{V} \right] = (C(x) - C_0) A; \, A = \text{constant}.$$

At the opening of the cylinder $x = 0$ and $C(x) = C_i$ so we may replace the arbitrary constant A in our solution and obtain

$$C(x) = C_0 + (C_i - C_0) \exp \left[- \frac{k 2\pi a}{V} x \right]. \tag{8.11}$$

From this equation it is seen that as x gets larger, the concentration of the fluid in the cylinder approaches that of the fluid on the other side of its walls.

At the end of the vessel $(x = l)$ the concentration C_l of the fluid is given by

$$C_l = (C_i - C_0) \exp \left[- \frac{k 2\pi a}{V} l \right] + C_0$$

The total change in concentration of the fluid as it passes along the whole length of the tube is given by

$$\Delta C = C_l - C_0$$

$$= (C_l - C_0)\left(\exp\left[-\frac{k2\pi a}{V}l\right] - 1\right).$$

In expression (8.11) it is seen that whatever the value of x, $C(x)$ will never exactly equal C_0 but will get very close to this value as x increases. In addition the value of $k2\pi a/V$ will determine how closely the exponential term approaches zero for a particular value of x. Thus varying the radius of the tube or the rate of flow of the solution through the tube will affect the concentration of the fluid emerging from the tube. Increasing the rate of flow will produce a less concentrated solution. Increasing the radius will produce a more concentrated solution.

8.6   DERIVATION OF GROWTH CURVES

If a population is allowed to grow in an environment in which space, food, water and all other necessities are unlimited, then the rate of growth is proportional to the population size. At the same time the rate of decrease in population due to death of its members is also proportional to the size of the population. We can set up a differential equation

$$\frac{dN}{dt} = bN - dN = r_m N \tag{8.12}$$

where N is the size of the population at time t, dN/dt is the rate of population growth, b a constant representing the birth rate, d a constant representing the death rate and r_m is a constant known as the innate capacity for increase.

If N_0 is the size of the population at time $t = 0$, the solution of the differential equation (8.12) is

$$N = N_0 \exp[r_m t]. \tag{8.13}$$

In practice, the environment is such that the growth of a population is limited. It will reach a state where the addition of more individuals will have the effect of decreasing the value of r_m and this will be more severely decreased, the larger the value of N. We may include this restriction as a modification of equation (8.12) as follows:

$$\frac{dN}{dt} = (r_m - cN)N \tag{8.14}$$

where c is a constant. This equation satisfies the stipulation that r_m is more affected at the large values of N.

Before we solve this equation we note that for a particular value of N the expression $(r_m - cN)$ will be zero and the rate of growth of the population will be zero. This means that the population size will attain a limit N_{max}.

For $N = N_{max}$,

$$\frac{dN}{dt} = 0 \text{ and } r_m - cN_{max} = 0.$$

Therefore we can express c in terms of N_{max} and r_m:

$$c = \frac{r_m}{N_{max}}.$$

Our differential equation will now take the form

$$\frac{dN}{dt} = r_m N \left\{ 1 - \frac{N}{N_{max}} \right\}.$$

This equation is inverted prior to integration so that

$$t = \frac{N_{max}}{r_m} \int \frac{dN}{N(N_{max} - N)}. \tag{8.15}$$

The integrand is split into partial fractions:

$$\frac{1}{N(N_{max} - N)} = \frac{A}{N} + \frac{B}{(N_{max} - N)}.$$

We cross multiply and obtain

$$1 = A(N_{max} - N) + BN,$$

and put $N = N_{max}$ to give

$$B = \frac{1}{N_{max}};$$

and put $N = 0$ to give

$$A = \frac{1}{N_{max}}.$$

Thus

$$\frac{1}{N(N_{max} - N)} = \frac{1}{N_{max}} \left\{ \frac{1}{N} + \frac{1}{(N_{max} - N)} \right\}.$$

This expression is substituted into the integral of expression 8.15 and we obtain

$$t = \frac{1}{r_m} \int \frac{dN}{N} + \frac{1}{r_m} \int \frac{dN}{N_{max} - N}$$

and after integration

$$t = \frac{1}{r_m} \{\ln N - \ln(N_{max} - N) + C\}. \qquad (8.16)$$

When $t = 0$, $N = N_0$ so we can obtain a value of the constant in terms of the initial population size N_0 and the final population size N_{max}:

$$0 = \frac{1}{r_m} \ln N_0 - \frac{1}{r_m} \ln(N_{max} - N_0) + C,$$

$$C = \frac{1}{r_m} \ln \left\{ \frac{N_{max} - N_0}{N_0} \right\}$$

We insert this value into (8.16), rearrange the equation, and obtain:

$$\exp[-r_m t] = \frac{N_0(N_{max} - N)}{N(N_{max} - N_0)}.$$

This is rearranged in order to obtain an expression for N:

$$N = \frac{N_{max}}{1 + b \exp[-r_m t]}, \qquad (8.17)$$

where

$$b = \frac{N_{max} - N_0}{N_0} \qquad (8.18)$$

Expression (8.17) is in fact the Logistic growth curve discussed in Chapter 7.

8.7   SEPARABLE VARIABLES

If our differential equation can be expressed in the form

$$\frac{dy}{dx} = \frac{f(x)}{F(y)} \qquad (8.19)$$

then the solution is obtained by first cross multiplying to obtain

$$F(y) \cdot \frac{dy}{dx} = f(x)$$

and then integrating with respect to x:

$$\int F(y) \cdot \frac{dy}{dx} dx = \int f(x) dx$$

so that our differential equation is solved by performing two integrations

$$\int F(y) dy = \int f(x) dx. \qquad (8.20)$$

Example I

Solve

$$x^2 \frac{dy}{dx} + y^2 = 1$$

We first of all put the equation in the form

$$\frac{dy}{dx} = \frac{1-y^2}{x^2}.$$

Thus

$$F(y) = \frac{1}{1-y^2}$$

and

$$f(x) = \frac{1}{x^2}.$$

These functions are inserted into expression 8.20:

$$\int \frac{dy}{1-y^2} = \int \frac{dx}{x^2}.$$

The left hand integral is first of all put in the form of partial fractions to facilitate the integration:

$$\int \frac{dy}{1-y^2} = \frac{1}{2}\int \frac{dy}{1+y} + \frac{1}{2}\int \frac{dy}{1-y}$$

$$= \frac{1}{2}\ln\frac{(1+y)}{(1-y)} + C.$$

The other integral is simply

$$\int \frac{dx}{x^2} = -\frac{1}{x} + C.$$

The solution of our differential equation is thus

$$\ln\frac{(1-y)}{(1+y)} = \frac{2}{x} + C.$$

or

$$\frac{1-y}{1+y} = \exp\left[\frac{2}{x} + C\right]$$

$$= A \exp\left[\frac{2}{x}\right], \ (A = \text{constant}).$$

If we rearrange this expression as an expression for y we obtain:

$$y = \frac{1 - A \exp[2/x]}{1 + A \exp[2/x]}.$$

Example II

Suppose that it is found in the early development of a species of plant that the rate of increase in height varies inversely as the cube of the age and directly as the height. Set up a differential equation to obtain a relationship between the age of the plant and its height.

First of all let

$$y = f(t) = \text{height of the plant,}$$
$$t = \text{age of the plant,}$$

$$\frac{dy}{dx} = \text{rate of increase in height.}$$

From the information given we may write

$$\frac{dy}{dt} = A\left(\frac{y}{t^3}\right)$$

The introduction of an unknown constant A is necessary because its value is not given to us. Its value will in fact be a peculiarity of the plant that is being studied. It is not to be confused with the arbitrary constant that will arise in the solution of the differential equation.

We obtain a solution by putting the equation in the form

$$\int \frac{dy}{y} = A \int \frac{dt}{t^3},$$

which, after integration, becomes

$$\ln(y + C) = -\frac{A}{2t^2}, \quad (C = \text{constant}).$$

or

$$y + C = \exp[-A/2t^2].$$

The constant C could be determined if we knew certain other information about the growth of the plant. Suppose, for instance, that the plant starts off as a seed and, after an initial period of time τ, the plant emerges from the surface of the ground so that

$$\text{when } t = \tau, \ y = 0,$$

our constant then has the value

$$C = \exp[-A/2\tau^2],$$

and our solution is

$$y = \exp[-A/2t^2] - \exp[-A/2\tau^2].$$

If this equation for its growth holds throughout the life of the plant, it is seen that, as t increases, the value of $\exp[-A/2t^2]$ in the above equation approaches the value unity but theoretically only after an infinite time.

We can thus have a value of the maximum height of the plant

$$y_{\text{max}} = 1 - \exp[-A/2\tau^2].$$

Exercises 8.7

Solve the following differential equations:

(1) $(y + a)\dfrac{dy}{dx} = (x + a)$

(2) $xy\dfrac{dy}{dx} = 1 + y^2$

(3) $\dfrac{dy}{dx} = x(y + b)$

(4) $x\dfrac{dy}{dx} = xy + y.$

8.8   HOMOGENEOUS EQUATIONS

Equations of the form

$$\frac{dy}{dx} = f(y/x) \tag{8.21}$$

are known as *homogeneous equations.* They may be solved by making the substitution.

$$z = \frac{y}{x} \tag{8.22}$$

so that

$$\frac{dz}{dx} = \frac{1}{x^2}\left\{x\frac{dy}{dx} - y\right\}$$

$$= \frac{1}{x^2}\left\{x\frac{dy}{dx} - xz\right\}$$

$$= \frac{1}{x}\left\{\frac{dy}{dx} - z\right\}$$

and

$$\frac{dy}{dx} = x\frac{dz}{dx} + z. \tag{8.23}$$

If the substitutions (8.22) and (8.23) are inserted into our differential equation (8.21) we obtain

$$x\frac{dz}{dx} + z = f(z).$$ (8.24)

The equation thus takes the form of a function in z.

Example I

Solve the differential equation

$$(x^2 - xy)\frac{dy}{dx} = xy + y^2$$

We divide throughout by x^2:

$$\left(1 - \frac{y}{x}\right)\frac{dy}{dx} = \frac{y}{x} + \frac{y^2}{x^2}\ .$$

We make the substitutions for y and dy/dx and the equation takes the form

$$(1 - z)\left(x\frac{dz}{dx} + z\right) = (z + z^2)$$

or

$$x\frac{dz}{dx} = \frac{2z^2}{1 - z}$$

Before we integrate we put the equation in the form

$$\frac{1 - z}{z^2}\frac{dz}{dx} = \frac{2}{x}\ .$$

We now integrate with respect to x:

$$\int\frac{1 - z}{z^2}dz = 2\int\frac{dx}{x}$$

and obtain

$$\ln A + \frac{1}{z} + \ln z = -2 \ln x; (A \text{ is a constant}).$$

The solution is rearranged to take the form

$$-\frac{1}{z} = \ln(zx^2 A).$$

We take exponentials of each side and obtain

$$\exp[-1/z] = Ax^2 z.$$

We finally replace z by y/x and the solution to our equation is

$$\exp[-x/y] = Axy.$$

Sometimes an equation is not recognisable as homogeneous until a preliminary substitution has been made. Suppose we have an equation of the form:

$$\frac{dy}{dx} = \frac{Ax + By + C}{ax + by + c} \tag{8.25}$$

where A, B, C, a, b, and c are constants.

We solve this by making the substitutions

$$Y = Ax + By + C; \tag{8.26}$$

$$\frac{dY}{dx} = A + B\frac{dy}{dx}, \tag{8.27}$$

and

$$X = ax + by + c; \tag{8.28}$$

$$\frac{dX}{dx} = a + b\frac{dy}{dx}. \tag{8.29}$$

From (8.27) and (8.29) we obtain

$$\frac{dY}{dX} = \frac{A + B\dfrac{dy}{dx}}{a + b\dfrac{dy}{dx}} \tag{8.30}$$

But if we make the substitutions (8.26) and (8.28) in our original equation (8.25), then we obtain

$$\frac{dy}{dx} = \frac{Y}{X}$$

which we substitute into (8.30) so that it takes the form

$$\frac{dY}{dX} = \frac{A + B\dfrac{Y}{X}}{a + b\dfrac{Y}{X}}.$$

(8.31)

This is a homogeneous equation which is soluble provided that A/B is not equal to a/b. If this is the case the single substitution

$$X = ax + by + c$$

is sufficient to bring the equation to a soluble form.

Example II

Solve

$$\frac{dy}{dx} = \frac{2x - 2y + 1}{2x + y - 1}$$

We make the substitutions

$$Y = 2x - 2y + 1; \frac{dY}{dx} = 2 - 2\frac{dy}{dx},$$

and

$$X = 2x + y - 1; \frac{dX}{dx} = 2 + \frac{dy}{dx}$$

so that

$$\frac{dY}{dX} = \frac{2 - 2(Y/X)}{2 + (Y/X)}.$$

This is a homogeneous equation so we make the substitution

$$z = \frac{Y}{X}; \frac{dY}{dX} = X\frac{dz}{dX} + z$$

and obtain

$$X\frac{dz}{dX} + z = \frac{2 - 2z}{2 + z}$$

which can be put in the form

$$\frac{2+z}{2-4z-z^2}\frac{dz}{dX}=\frac{1}{X}$$

so that we can integrate with respect to X:

$$\int\frac{2+z}{2-4z-z^2}dz=\int\frac{dX}{X}. \qquad (8.32)$$

The left hand integral is of the form which is solved by making the substitution

$$W=2-4z-z^2;\frac{dW}{dz}=-4-2z \qquad (8.33)$$

so that

$$\int\frac{2+z}{2-4z-z^2}dz=-\frac{1}{2}\int\frac{dW}{W}=-\frac{1}{2}\ln(W)+C.$$

The right hand integral of expression (8.32) is simply

$$\int\frac{dX}{X}=\ln X+C.$$

The solution of our differential equation is thus

$$\ln W+2\ln X=\ln c \ (c=\text{constant}).$$

By getting rid of the logarithms we obtain

$$W X^2=c.$$

We now insert the appropriate functions for W and z to obtain our solution in terms of X and Y:

$$2X^2-4XY-Y^2=c.$$

The original expressions for X and Y are now inserted into the solution

$$2(2x+y-1)^2-4(2x+y-1)(2x-2y+1)$$
$$-(2x-2y+1)^2=c.$$

The brackets are multiplied out and like terms added to obtain

$$-12x^2+24xy+6y^2-12x-12y+5=c$$

Finally we lump $c - 5$ together as a constant and tidy up the expression by dividing throughout by -6.

$$2x^2 - 4xy - y^2 + 2x + 2y = c'$$

where

$$c' = \frac{5 - c}{6}.$$

Example III

Solve the differential equation

$$\frac{dy}{dx} = \frac{2x - 2y + 6}{x - y}.$$

In this example we do not put the equation in the form of expression (8.31) because here the values of the ratios A/a and B/b are equal. Instead we make the substitution

$$X = x - y; \frac{dX}{dx} = 1 - \frac{dy}{dx},$$

into our equation so that it becomes

$$1 - \frac{dX}{dx} = \frac{2X + 6}{X}.$$

This may be simplified

$$\frac{dX}{dx} = -\frac{X + 6}{X}$$

and then put in the form

$$\frac{X}{X + 6}\frac{dX}{dx} = -1$$

so that it may be integrated with respect to x:

$$\int \frac{X}{X + 6}dX = -\int dx.$$

The right hand integral is simply equal to x + constant and the left hand integral is

$$\int \frac{X}{X+6}\,dX = \int \left(1 - \frac{6}{X+6}\right)dX$$

$$= X - 6\ln(X+6) + C.$$

$$= (x-y) - 6\ln(x-y+6) + C.$$

The complete solution of the differential equation is thus:

$$2x - y - 6\ln(x - y + 6) = C.$$

Exercises 8.8

(1) $(x + y + 1)\dfrac{dy}{dx} = (x - y + 1)$ (2) $(x + y)\dfrac{dy}{dx} = (x + y + 3)$

(3) $(2y - x)\dfrac{dy}{dx} = (2x + y)$ (4) $\dfrac{dy}{dx} = \dfrac{(x - y + 2)}{(x + y - 2)}$

(5) In the study of servo-mechanisms and their application to biological systems, the following differential equation sometimes arises

$$y\frac{dy}{dx} + 2Ky + x = 0$$

where K is a constant. Solve this equation for the case where $K^2 > 1$.

8.9   LINEAR EQUATIONS

Equations of the form

$$\frac{dy}{dx} = yP(x) + Q(x) \tag{8.34}$$

where P and Q are functions of x only, are known as *linear* differential equations of the first order.

To solve such equations we put

$$w = \int P(x)\,dx \text{ with the arbitrary constant zero.}$$

and

$$z = y \exp [-w]$$

so that

$$\frac{dz}{dx} = \exp [-w] \frac{dy}{dx} - y \frac{dw}{dx} \exp [-w]$$

$$= \left\{ \frac{dy}{dx} - yP(x) \right\} \exp [-w]$$

$$= Q(x) \exp [-w]$$

This expression is now integrated with respect to x:

$$z = \int Q(x) \exp [-w] \, dx.$$

But

$$z = y \exp [-w],$$

so

$$y = e^w \int Q(x) \exp [-w] \, dx$$

$$= \exp \left[\int P(x) \, dx \right] \int Q(x) \exp \left[- \int P(x) \, dx \right] dx. \qquad (8.35)$$

The equation is solved by putting the appropriate functions into equation (8.35) above. It can be proved that this is indeed a solution to our differential equation of (8.34) simply by differentiating (8.35) with respect to x.

Certain equations may be put into the linear form by making an appropriate substitution. A type of differential equation known as the *Bernoulli Equation* has the form

$$\frac{dy}{dx} = yP(x) + y^n Q(x) \qquad (8.36)$$

and by making the substitution

$$z = y^{1-n}$$

so that

$$\frac{dz}{dx} = (1 - n) y^{-n} \frac{dy}{dx}$$

or

$$\frac{dy}{dx} = \frac{y^n}{(1-n)} \cdot \frac{dz}{dx}$$

so that equation (8.36) becomes

$$\frac{y^n}{(1-n)} \frac{dz}{dx} = yP(x) + y^n Q(x)$$

or

$$\frac{dz}{dx} = (1-n)\{y^{1-n}P(x) + Q(x)\}$$

$$= (1-n)\{zP(x) + Q(x)\}.$$

We have transformed our equation into a linear differential equation of the form of 8.34.

Exercises 8.9

Solve the following differential equations:

(1) $xy\dfrac{dy}{dx} = x^2 + y^2$ $\qquad\qquad$ (2) $x\dfrac{dy}{dx} = (x^2 - y)$

(3) $x^2\dfrac{dy}{dx} = 2y^2 - xy$ $\qquad\qquad$ (4) $\dfrac{dy}{dx} = x^2 - \dfrac{y}{x}$

(5) $(1 + x^2)\dfrac{dy}{dx} = x(1 - y)$ $\qquad\qquad$ (6) $x\dfrac{dy}{dx} - x + y = 0.$

8.10   SECOND ORDER DIFFERENTIAL EQUATIONS

These equations contain the second derivative; they may contain the first derivative as well, but third and higher derivatives are not present. Only one type of second order differential equation will be considered here and this is

$$\frac{d^2 y}{dx^2} = f(x).$$

This requires two integrations with respect to x. Firstly

$$\frac{dy}{dx} = \int f(x)\,dx + A,$$

where A is an arbitrary constant. After the second integration we obtain

$$y = \int \left\{ \int f(x) \, dx \right\} dx + Ax + B$$

where B is an arbitrary constant. In this case, as in the case of all second order differential equations, two arbitrary constants are involved in the answer.

Example I

A particle moves in a straight line with constant acceleration k. Derive an expression for its position at any time t given that when

$$t = 0, \; dy/dt = 0 \text{ and } y = 0.$$

Our differential equation is

$$\frac{d^2 y}{dt^2} = k$$

After the first integration we obtain

$$\frac{dy}{dt} = kt + A.$$

Since we are given that when $t = 0$, $dy/dt = 0$ the constant A has zero value.

After the second integration we obtain

$$y = \tfrac{1}{2}(kt^2) + B.$$

Since when $t = 0$, $y = 0$, the constant B is also zero. The solution of our differential equation is thus

$$y = \tfrac{1}{2}(kt^2).$$

Exercises 8.10

Solve the following differential equations:

(1) $\dfrac{d^2 y}{dx^2} = x^n$

(2) $x \dfrac{d^2 y}{dx^2} = A$

(3) $\dfrac{d^2 y}{dx^2} = \exp x$

(4) $\dfrac{d^2 y}{dx^2} = x \exp x$

(5) $x^2 \dfrac{d^2 y}{dx^2} = 1$

(6) $\dfrac{d^2 y}{dx^2} = \ln x.$

SOLUTIONS TO EXERCISES

Exercises 1.1 (Page 6)

(1) 15552   (2) 6   (3) 1

(4) $x^3 y^7$ (5) $3x^4 y^3 - x^2 y^2$

(6) 16   (7) $2y^3/x^3$ (8) $4a^2 b^2 d^3$

(9) $a^4(b + c)$ (10) (i) 0.0128 or 8/625

(ii) 0.375 or 3/8.

(11) The size of the population after 10 hours in the first medium will be

$$1000.2^{10}.$$

0.1% of these, i.e. 2^{10} are now placed in the second culture medium for 10 generation times. At the end of this time the number of cells in the second medium will be

$$2^{10} . 2^{10} = 2^{20}$$
$$= 1,048,576.$$

(12) The rate of flow will increase by 21.6% (to three figures).

Exercises 1.2 (Page 10)

(1) 243   (2) 1   (3) $\frac{1}{3}$ (4) 2

(5) 1   (6) 49   (7) $\frac{1}{8}$ (8) $1/2a^2 b.$

Exercises 1.3 (Page 16)

(2) 0

(3a) Put $y = 2^x$ so that

$$2^{2x} - 12.2^x + 32 = 0$$

becomes

$$y^2 - 12y + 32 = 0.$$

When this is factorised we have

$$(y - 8)(y - 4) = 0$$

and we obtain two solutions;

$$\text{(i) } y = 2^x = 4 = 2^2$$
$$\text{and therefore } x = 2$$
$$\text{(ii) } y = 2^x = 8 = 2^3$$
$$\text{and therefore } x = 3.$$

(3b) Put $y = 3^{2x}$ so that

$$3^{4x} - 2.3^{2x} + 1 = 0$$

becomes

$$y^2 - 2y + 1 = 0.$$

When this is factorised we have

$$(y - 1)^2 = 0$$

and we obtain the single solution

$$y = 3^{2x} = 1,$$

therefore

$$x = 0.$$

(4) 0.699   (5) $x = \tfrac{1}{3}$ (6) $x = 4$.

Exercises 1.4 (Page 19)

(1) 3.8454	(2) 0.8454	(3) 2.8454	(4) $\bar{1}.8454$
(5) 1.0145	(6) 3.0145	(7) $\bar{1}.0145$	(8) 4.0149
(9) 3.8999	(10) 1.9000	(11) 0.9003	(12) 2.9996
(13) $\bar{4}.0000$	(14) $\bar{3}.3222$	(15) $\bar{1}.9156$	(16) 2.8567.

Exercises 1.5 (Page 21)

(1) 1.500   (2) 1.499   (3) 1.496   (4) 9.940

(5) 99.60   (6) 100.2   (7) 0.1379   (8) 0.001122

(9) 3.236   (10) $8.129 \ 10^7$ (11) 0.2139   (12) 105.0.

Exercises 1.6 (Page 25)

(1)

   (a) 15.4   (b) 15.1   (c) 153.0   (d) 0.165   (e) 0.253

(2)

   (a) 5.49 inches   (b) 110 cu.mm

(3) 0.752

(4) 16.4.

Exercises 1.7 (Page 30)

(1)

   (a) 2.3410   (b) 5.2076   (c) 1.7696

   (d) 9.1433   (e) 5.5086   (f) 4.5376.

(2)

   (a) 0.000005012   (b) $1.585 \ 10^{-12}$

   (c) 10^{-7} (d) 0.001259

   (e) $1.259 \ 10^{-7}$ (f) $1.585 \ 10^{-6}$

   (g) $7.943 \ 10^{-11}$ (h) $6.310 \ 10^{-7}$

(3) The concentration increases by $9.27 \ 10^{-13}$ g.moles per litre.

(4)

   (a) 6.4010   (b) 7.12

(c) 19.05   (d) 6.9651

(e) 7.942   (f) 5.012

(g) 7.3010   (h) 7.942

(i) 5.6029

(5) Amount of acid added to change the pH to $6.0 = 2.44\ 10^{-6}$ g.mole per litre. Amount of hydrogen ion removed to change the pH from 6.1 to $6.3 = 3.7\ 10^{-6}$ g.mole per litre. When the amount of buffer is reduced the amount of acid needed to change the pH from 6.1 to 6.0 is $2.30\ 10^{-7}$ g. mole per litre.

Exercises 1.8 (Page 32)

(1)
  (a) 2.04   (b) 2.10   (c) 1.89   (d) 1.83   (e) 1.69

(All these answers are expressed in square metres).

(2) 15.6

Exercises 1.9 (Page 34)

(a) 12.2   (b) 2.46 cm.   (c) $7.71\ 10^{-7}$

Exercises 2.1 (Page 37)

(1)
  (a) 70   (b) $2\frac{41}{42}$ (c) $\frac{1}{30}$ (d) $\frac{5}{24}$ (e) 5

(2)
  (a) $\dfrac{391}{720}$ (b) $\dfrac{5}{8}$ (c) $\dfrac{n(n-1)(n-2)}{n+2}$

(3)
  (a) $\dfrac{n!}{(n-r)!}$ (b) $\dfrac{(n+r)!}{(n-1)!}$

(4)

(a)
$$1 \times 3 \times 5 \times 7 \ldots (2n - 1)$$

$$= \frac{1 \times 2 \times 3 \times 4 \times 5 \times 6 \times 7 \times 8 \ldots (2n - 1)}{2 \times 4 \times 6 \times 8 \ldots (2n - 2)}$$

$$= \frac{(2n - 1)!}{2^{n-1} (1 \times 2 \times 3 \times 4 \ldots n - 1)}$$

$$= \frac{(2n - 1)!}{2^{n-1}(n - 1)!}$$

$$= \frac{(2n - 1)! \, 2n}{2^{n}(n - 1)! \, n}$$

$$= \frac{(2n)!}{2^{n}(n)!}$$

(b)
$$\frac{11}{2} \times \frac{9}{2} \times \frac{7}{2} \times \frac{5}{2} \times \frac{3}{2} \times \frac{1}{2}$$

$$= \frac{11 \times 10 \times 9 \times 8 \ldots 3 \times 2 \times 1}{2^{6} \, 10 \times 8 \times 6 \times 4 \times 2}$$

$$= \frac{11!}{2^{11} \, 5 \times 4 \times 3 \times 2 \times 1}$$

$$= \frac{11! \, 12}{2^{12} \, 6 \times 5 \times 4 \times 3 \times 2 \times 1}$$

$$= \frac{12!}{2^{12} \, 6!}$$

Exercises 2.2 (Page 40)

(1)
   (a) 1.649   (b) 1.105   (c) 0.1353

(2)
   (a)
$$e^{4} = e^{3} \cdot e$$

$$= 20.09 \times 2.718$$

$$\simeq \underline{54.6}$$

(b)
$$e^{-2} = \frac{1}{e^2}$$

$$= \frac{1}{(2.718)^2}$$

$$\simeq \underline{0.135}$$

(c)
$$e^6 = (e^3)^2$$

$$= (20.09)^2$$

$$\simeq \underline{404}$$

(d)
$$e^5 = \frac{e^6}{e} \simeq \underline{148}.$$

Exercises 2.3 (Page 43)

(1)

(a) 4.3733	(b) −0.6990	(c) −5.4262	(d) 7.1680
(e) 0.9980	(f) 4.2878	(g) 13.6991	(h) −4.5008
(i) 3.7610	(j) 0.2730	(k) 1.1448	(l) −11.5129
(m) 4.7194			

(2)

(a) 22.19	(b) 0.01499	(c) 0.6550	(d) 3.350
(e) $5.264 \ 10^7$	(f) 1.260	(g) 134.2	(h) 2.163

Exercises 2.4 (Page 46)

At time $t = 0.1\tau$ the first population will have a size

$$A \exp [0.1] = 1.1052 \, A,$$

and the second population will have a size

$$\tfrac{1}{2} A \exp [0.2] = 0.6107 \, A.$$

At time $t = 0.5\tau$ the first population will have a size

$$A \exp [0.5] = 1.6487\ A,$$

and the second population will have a size

$$\tfrac{1}{2} A \exp [1.0] = 1.3591\ A.$$

At time $t = \tau$ the first population will have a size

$$A \exp [1.0] = 2.7183\ A,$$

and the second population will have a size

$$\tfrac{1}{2} A \exp [2.0] = 3.6946.$$

Let the populations have equal size at time $t = T$. In this situation we have

$$A \exp [T/\tau] = \tfrac{1}{2} A \exp [2T/\tau]$$

or

$$2 \exp [T/\tau] = \exp [2T/\tau].$$

We take natural logarithms of both sides and obtain

$$\ln 2 + T/\tau = 2T/\tau$$

or

$$T = \tau \ln 2$$
$$= 0.6931\ \tau.$$

Exercises 2.5 (Page 49)

(1)

The proportion P of undecayed matter remaining after a time t is given by

$$P = \frac{C}{C_0} = \exp \left[-\frac{0.693}{14.2}\, t \right]$$

This equation is rearranged to the form

$$t = -\frac{14.2}{0.693} \ln P$$

(a) When 1% of the activity is lost $P = 0.99$ so that

$$t = -\frac{14.2}{0.693} \ln P$$

$$= -\frac{14.2}{0.693}(-0.0101)$$

$$= 0.207 \text{ days.}$$

(b) When 90% of the activity is lost $P = 0.1$ so that

$$t = -\frac{14.2}{0.693}(\ln P)$$

$$= -\frac{14.2}{0.693}(-2.3026)$$

$$= 47.2 \text{ days.}$$

(2) We must first of all determine the proportion P of undecayed material remaining after one day.

$$P = \frac{C}{C_0} = \exp\left[-\frac{0.693}{14.2}\right]$$

$$= \exp[-0.04880]$$

$$= 0.9524.$$

Therefore just over 95% of the material remains after one day. Any sample that is taken must therefore be corrected by a factor

$$\frac{1}{0.9524} = 1.050$$

to account for the amount of material that has decayed.

Exercises 3.3 (Page 59)

(1)
    (a) 4.561   (b) 9.22
    (c) 1.166   (d) 5

(2)
    3.89 mm.

Exercises 3.4 (Page 64)

(1)
    (a) 0.3564   (b) 0.2214   (c) 0.6536
    (d) −4.4922   (e) −0.1492   (f) −0.1763
    (g) 0.6536   (h) −5.3955

(2) There is no unique answer as to which angle has a particular tangent because

$$\tan \theta = \tan (\theta + n.180); [\theta \text{ extends from } 0° - 180°]$$

where n is any integer or zero. The answers are given here for the cases where $n = 0$.

    (a) $86° 48'$ (b) $30° 31'$

    (c) $124° 59'$ (d) $39° 0'$

    (e) $119° 26'$ (f) $19° 43'$

(3)

    (a) -4.043 (b) 1.5

    (c) 1 (d) 0.7976

    (e) 0

Exercises 3.5 (Page 69)

(1)

    (a) $y = x - 1$ (b) $(a + b)y = (b - a)x + a^2 + b^2$

    (c) $3y = x + 2$ (d) $x + y - 4 = 0$

    (e) $2y - 6 + 3x = 0$ (f) $y = x - 1$

(2)

    (a) $y = 2x + 3$ (b) $y = x + 1$

    (c) $y = 3x + 3$ (d) $y + x = 1$

(3) $y = mx - md$

Exercises 3.7 (Page 75)

(a) Intercept 0.9, slope 0.385

(b) Intercept 5.0, slope $-\frac{1}{2}$

Exercises 3.9 (Page 79)

By the method of averages

$$\text{the slope} = -0.5125$$

$$\text{the intercept} = 4.52.$$

By the method of least squares

$$\text{the slope} = -0.524$$

$$\text{the intercept} = 4.56.$$

Exercises 3.10 (Page 81)

(2) $x = -1.903y + 8.688.$

Exercises 3.11 (Page 82)

The value of r for the 8 pairs of values is 0.9986. From Table 3.IV it is seen that $r = 0.71$ for $n = 8$. Our calculated value is much better than this. The probability of a linear relationship is therefore greater than 95%.

Exercises 4.1 (Page 88)

(1) $y = 2.9 x^{1.5}$

(2) $y = 2.25 x^{-1}$

(3) $y = 2 x^{0.8}$

Exercises 4.4 (Page 99)

(1)

(a) $y^2 - 10y - 4x + 19 = 0$

(b) $y = \cos(x) + \sin(x + 9) + 5$

(c) $(x + 4)^2 = (y - 9)^2$

(2)

In order to transform the axes by rotation through $45°$, make the substitutions

$$x = x \cos 45 - y \sin 45;$$

$$y = x \sin 45 + y \cos 45.$$

But

$$\sin 45 = \cos 45 = \frac{1}{\sqrt{2}}$$

so we obtain the following answers:

(a) $xy = -\frac{1}{2}$

(b) $y = \frac{1}{\sqrt{2}}$

(c) $x^2 - y^2 = 4.$

Exercises 4.7 (Page 107)

(a) $A = 30, B = 0.24$

(b) $A = 8, B = -0.347$

Exercises 5.2 (Page 118)

(1) When $x = 0.5$ the slope of the tangent is 3.
When $x = 3$ the slope of the tangent is 8.
When $x = 10$ the slope of the tangent is 22.

(2) When $x = 0$ the slope of the tangent is 4.
When $x = 1$ the slope of the tangent is 6.
When $x = 2$ the slope of the tangent is 8.

(3) When $x = 1$ the slope of the tangent is 0.
When $x = 1.5$ the slope of the tangent is 1.
When $x = 0.8$ the slope of the tangent is -0.4.

Exercises 5.3 (Page 121)

At time $t + \tau$ the size of the population is

$$N(t + \tau) = 10^5 (1 + 2[t + \tau] + [t + \tau]^2).$$

The increase in size over the period of time $t \to t + \tau$ is

$$N(t + \tau) - N(t) = 10^5 (2\tau + 2t\tau + \tau^2)$$

The mean rate of increase over this period of time is

$$\frac{N(t + \tau) - N(t)}{\tau} = 10^5 (2 + 2t + \tau)$$
$$= 10^5 (2 + 2t)$$

as τ approaches zero.

When $t = \frac{1}{2}$ hour the rate of growth is 3.10^5 individuals per hour,
When $t = 2$ hours the rate of growth is 6.10^5 individuals per hour.

Exercises 5.4 (Page 123)

(1) 4 (2) 27 (3) 3 (4) -1.5 (5) $\dfrac{20}{3}$

(6) $3 b^2$ (7) 6 (8) -2 (9) 1.5 (10) -4.

Exercises 5.5 (Page 128)

(1) $f'(x) = 6x$, $f'(3) = 18$, $f'(0) = 0$.

(2) $3 \theta^2$

(3) 4

(4) $2x/5$

(5) $z'(t) = -(t + 2)^{-2}$,
    $z'(-4) = -\dfrac{1}{4}$

Exercises 5.6 (Page 131)

(1) $3x^2 + 6x + 4$

(2) $\dfrac{4(x + 1)}{x^2(x + 2)^2}$

(3) $\dfrac{2x}{5} + 4$

Exercises 5.7 (Page 136)

(1) $3x^2 + 1$

(2) $1 - 2x$

(3) $2x + 3$

(4) $-(x + 1)^{-2}$

(5) $3x^2 + 6x + 2$

(6) $3(x + 4)^2$

(7) $\dfrac{(x + 1)(x - 3)}{(x - 1)^2}$

(8) $(x + 3)(x - 1)^2(5x + 7)$

(9) $\dfrac{1 - 2x}{x^2(x - 1)^2}$

(10) $3(x - 1)^2$

(11) $4x(x^2 - 1)$

(12) $3(x - 1)(x - 3)$

(13) $2x + (a - b)$

(14) $(x + a)(3x + a - 2b)$

Exercises 5.9 (Page 145)

(1) $4x^3 + 6x$

(2) $3(x + 1)^2$

(3) $4x(x^2 + 2)^2(2x^2 + 1)$

(4) $-\frac{1}{2}x^{-2}$

(5) $-\frac{1}{3}x^{-4/3}$

(6) $\pi x^{\pi - 1}$

(7) $\frac{1}{2}(x + 1)^{-1/2}$

(8) $-2x\,e(x^2 + 1)^{-e-1}$

(9) $-(x^2 - 1)^{-3/2}$

(10) $-4(x - 1)^{-5}$

(11) $x^2(x - 1)(5x - 3)$

(12) $-5x^{-6}$

(13) $-9(3x - 2)^{-4}$

(14) $\frac{3}{2}x^{1/2}$

(15) $\frac{\pi}{5}(x + 2)^{\frac{\pi}{5} - 1}$

(16) $x^{-\frac{5}{6}}\left(\frac{1}{6} + x^{-1/2}\right)$

(17) $-(4x - 1)^{-3/2}\,x^{-1/2}$

(18) $\frac{3}{2}(x + 3)^{1/2}$

(19) $(x + 1)(x - 1)^e\left\{2 + \dfrac{e(x + 1)}{x - 1}\right\}$

(20) $-2(a + x)^{-3}$

(21) $\frac{1}{2}x^{-\frac{3}{2}}(x - 1)$

(22) $-\frac{1}{2}(x^3 + 2x + 1)^{-3/2}(3x^2 + 2)$

(23) $-\frac{1}{2}[x^{1/2}(x^{1/2} - 1)^2]^{-1}$

(24) $\frac{1}{2}(a + x)^{-1/2}(b - x)^{-3/2}(a + b)$

Exercises 5.10 (Page 148)

(1)

   (a) $(x + 5) e^x$

   (b) $5e^x - e^{-x}$

   (c) $e^x(1 + x + x^{-1} - x^{-2})$

   (d) $2x(1 + x) \exp [2x]$

   (e) $b \exp [a + bx]$

(2) Let A be the total amount of the drug in the circulation at any particular time t, then

$$A = V C(t).$$

The rate of change of amount of the substance in the circulation is

$$\frac{dA}{dt} = - V(G_1 k_1 \exp [-k_1 t] + G_2 k_2 \exp [-k_2 t])$$

(3) $\dfrac{A e}{\tau}$

Exercises 5.11 (Page 151)

(1) $6(2x + 4)^2$ (2) $-4(2x + 1)^{-3}$

(3) $8x + 6$ (4) $-2(1 - x)^2(2x + 1)$

(5) $2(2 - x) \dfrac{(1 + x)^2}{(1 - x)^2}$ (6) $2x + 7$

(7) $4x(x^2 + 1) - 2x(x^2 + 1)^{-2}$ (8) $-4x^3 \exp [-x^4]$

(9) $\frac{1}{2}(x^3 + 2x)^{-1/2}(3x^2 + 2) \exp [(x^3 + 2x)^{1/2}]$

Exercises 5.12 (Page 153)

(1) $- 3(2x - 1)^{-2}$ (2) $(x^2 + 4x + 6)(x + 2)^{-2}$

(3) $(x^3 - 3x^2)(x^2 - 1)^{-2}$ (4) $(12 - 3x^2)(x^2 + 4)^{-2}$

(5) $(6x - x^3)(3 - x)^{-2}$ (6) $(21 + 7x^2)(3 - x^2)^{-2}$

(7) $-2x(x^2-2)^{-2}$

(8) $(x^2-2x-4)(x-1)^{-2}$

(9) $-(3x+11)(x-1)^{-3}$

(10) $-4x(x^2-1)^{-11}(3x^2+1)^3(9x^2+11)$

(11) $16x(2x+1)(4x+1)^{-2}$

Exercises 5.13 (Page 156)

(1) $1/[3(y^2+1)]$

(2) $(y+a)^2(y^2+2ay)^{-1}$

(3) $(4y+9)^{-1}$

(4) $-(y+1)^2(2y^3+3y^2+1)^{-1}$

Exercises 5.14 (Page 159)

(1) $a^{x+2}\ln a$

(2) $2x(x^2+3)^{-1}$

(3) $(x\ln 10)^{-1}$

(4) $(3x^2+2)(x^3+2x)^{-1}$

(5) $6x(x^2+1)^{-1}$

(6) $e^x+1+\ln(x)$

(7) $\ln(x)$

(8) $e^x(x+2)+2/x$

(9) $e^x(x+1+x^{-1}-x^{-2})$

(10) $3\exp[3x]$

(11) $e^x(1+x)$

(12) $\ln(x)+[(x+1)/x]$

(13) $[2-\ln(x^2)]/x^2$

(14) $1/x$

(15) $\dfrac{1-x^2}{x(x^2+1)}$

Exercises 5.15 (Page 161)

(1) $-n\sin(nx)$

(2) $n\cos(nx)$

(3) $1+2\cos x$

(4) $3x^2\cos(x^3)$

(5) $2\sin(x)\cos(x)$

(6) $\sin(x)+\cos(x)$

(7) $\cos(x)\exp[\sin x]$

(8) $\dfrac{\cos x}{\sin x}$

(9) $\tfrac{1}{2}(\sin x)^{-1/2}\cos(x)$

(10) $-n\sin(x)\cos^{n-1}(x)$

(11) $2b\cos(x)[a+b\sin x]$

(12) $\cos^3 x-2\sin^2 x.\cos x$

Exercises 5.16 (Page 162)

(1) $\dfrac{2 \sin \theta}{\cos^2 \theta} (\cos^3 \theta + 1)$

(2) $\dfrac{2x - 1}{1 - 2x + 2x^2}$

(3) $2 \cos (2x) \cos (x) - \sin (2x) \sin (x)$

(4) $- 3 \sin (\pi + 3\theta)$

(5) $- \frac{2}{3} \exp [- 2x]$

(6) $\frac{1}{2} (\tan (\theta/2) + \tan^{-1} (\theta/2))$

$$= \frac{1}{2 \sin (\theta/2) \cos (\theta/2)}$$

(7) $3.4^{3z} \ln 4 + 3/z$

(8) $- \tan \theta$

(9) $(2 - y)(y + 1)^2 \exp [1 - y]$

(10) $\dfrac{\cos (3y) + 3y \sin (3y)}{\cos^2 (3y)}$

(11) $(0.2)^y \ln (0.2)$

(12) $\dfrac{1}{(p + 1)^2} - \dfrac{2}{(p + 2)^2}$

(13) 1

(14) $\exp [x + e^x]$

(15) Put $y = \log_{10}(e^x)$

(16) $- \sin \theta \cdot \exp [\cos \theta]$

$= \ln (10) \cdot \ln (e^x)$

$= \ln (10) \cdot x,$

so that

$$\frac{dy}{dx} = \ln (10).$$

Exercises 6.2 (Page 166)

(1) Minimum at $x = 3$

(2) Minimum at $x = 2$, maximum at $x = -2$

(3) Minimum at $x = \frac{2}{3}$, maximum at $x = -4$

(4) Minimum at $x = 3$, maximum at $x = -1$

Exercises 6.5 (Page 170)

(1)

   (a) $3 \cos x$, $-3 \sin x$

   (b) $- 4 \exp [-4x]$, $16 \exp [-4x]$

   (c) $1 - 1/x^2$, $2/x^3$

(d) $(1 - x) \exp [-x]$, $(x - 2) \exp [-x]$

(e) $3x(2 - x)$, $6(1 - x)$

(f) $\dfrac{2x}{(1 - x^2)^2}$, $\dfrac{2 - 10x^2}{(1 - x^2)^3}$

(g) $\sin \theta + \cos \theta$, $\cos \theta - \sin \theta$

(h) $3 \exp [3x]$, $9 \exp [3x]$

(2)

(a) Maximum at $x = -1$

(b) Maximum at $x = 0$, minima at $x = 2$ and $x = -2$

(c) If $y = \exp [x] - \exp [2x]$

then $\dfrac{dy}{dx} = \exp [x] - 2 \exp [2x]$

$= 0$ for a maximum or a minimum, that is when $\exp [x] = \frac{1}{2}$.

$$\frac{d^2 y}{dx^2} = \exp [x] - 4 \exp [2x]$$

$$= -\tfrac{1}{2} \text{ when } \exp [x] = \tfrac{1}{2}$$

Therefore a maximum occurs where

$\exp [x] = \frac{1}{2}$

or $x = \ln (\frac{1}{2})$.

(d) Maximum at $x = 1$.

Exercises 6.6 (Page 175)

(1) Let the box have a base-length B and height H. The area of material required is thus

$$A = B^2 + 4HB$$

and we also have the relationship

$$V = B^2 H,$$

$$\text{or } H = \frac{V}{B^2}$$

so that

$$A = B^2 + \frac{4V}{B}.$$

We now differentiate

$$\frac{dA}{dB} = 2B + \frac{4V}{B^2}$$

$$= 0 \text{ for a maximum or a minimum, i.e. when}$$

$$2B^3 = 4V$$

For a minimum

$$\frac{d^2A}{dB^2} = 2 + \frac{8V}{B^3}$$

must be positive when $dA/dB = 0$ and this condition is satisfied when $2B^3 = 4V$.

The area is thus a minimum when $2B^3 = 4V$, or $B = (2V)^{1/3}$

The minimum value of A is

$$A_{min} = B^2 + 2B^2 = 3B^2,$$

or, in terms of the given volume V,

$$A_{min} = 3(2V)^{2/3}.$$

(2) Length of wire forming the triangle

$$= \frac{3\sqrt{3}L}{3\sqrt{3} + \pi}.$$

Length of wire forming the circle

$$= \frac{\pi L}{3\sqrt{3} + \pi}.$$

(3) Initially, let the hunter be at a point A and the lion at a point B one quarter of a mile due south of A. Let the hunter move in a certain direction making an angle with AB. After a time t the line will reach a point C due west of B. The hunter will arrive at a point D at time t. For the particular chosen direction he will be at his closest to the lion when the points ADC lie on a straight line. We now need to determine the minimum distance CD and the angle that the hunter must travel in order to achieve it.

From Pythagorus theorem

$$AB^2 + BC^2 = AC^2.$$

But

$$AC = AD + DC,$$

10

so

$$AB^2 + BC^2 = [AD + DC]^2.$$

Let the lion and the hunter be nearest to each other at time T when the lion will have travelled a distance $AC = 20T$ miles and the hunter will have travelled a distance $AD = 18T$ miles, and

$$(\tfrac{1}{4})^2 + (20T)^2 = [18T + DC]^2$$

or

$$DC = [(\tfrac{1}{4})^2 + (20T)^2]^{1/2} - (18T)$$

If we differentiate this with respect to T and equate to zero we obtain

$$T = \frac{9}{80\sqrt{19}} \text{ hr.}$$

and it can be shown that the second derivative takes on a positive value for this value of T thus proving that it gives a minimum value of DC. This is calculated to be

$$\frac{\sqrt{19}}{40} \text{ miles} \qquad\qquad \text{or}$$

$$192 \text{ yards.}$$

The lion will have travelled a distance $BC = 20T$

$$= \frac{9}{4\sqrt{19}} \text{ miles}$$

The angle CAB will therefore have a tangent

$$= \frac{CB}{AB} = \frac{9}{\sqrt{19}}$$

and $CAB = 64° \ 10'$.

Exercises 6.8 (Page 181)

(1)
    (a) $v = -Ak \exp[-kt]$; $f = Ak^2 \exp[-kt]$

    (b) $v = A/t$; $f = -A/t^2$

  (c) $v = 4; f = 0$

  (d) $v = u + gt; f = g.$

(2)

  (a) 0

  (b) $- 3A$

Exercises 7.3 (Page 185)

(1) $C - x^{-4}/4$

(2) $\dfrac{x^{0.9}}{0.9} + C$

(3) $x + C$

(4) $\dfrac{x^{3.5}}{3.5} + C$

Exercises 7.7 (Page 189)

(1) $4 \sin (x) + C$

(2) $4 x^{3/2} + C$

(3) $\dfrac{8x^3}{3} + C$

(4) $\dfrac{3x^{1/3}}{2} + C$

Exercises 7.9 (Page 194)

(1) $\dfrac{1}{a} \ln (ax + b) + C$

(2) $\dfrac{(3x + 2)^6}{18} + C$

(3) $- \tfrac{1}{6} \cos (6t - \tfrac{1}{2}\pi) + C$

(4) $- \tfrac{2}{3}(3 - x)^{3/2} + C$

(5) $- \tfrac{1}{2}(t + 3)^{-2} + C$

Exercises 7.10 (Page 202)

(1) 3.5   (2) 1.983   (3) $\frac{1}{3}$

(4) 0.6931 (= ln 2)

(5) 3672   (6) $\frac{8}{3}$

(7) 14$\frac{2}{3}$.

Exercises 7.11 (Page 204)

(1) 65$\frac{1}{3}$ (2) 16$\frac{2}{3}$ (3) 48$\frac{2}{3}$

(4) $2 - \sqrt{2}$ (5) 4.793   (6) 2

(7) 70.98   (8) $\frac{4}{9}$ (9) $\sqrt{2}$

(10) 63   (11) $\frac{15}{4}$ (12) 6.429

(13) $2\sqrt{2}$ (14) $1/(8\pi)$ (15) 1181.25

(16) 0.333   (17) $(4\sqrt{2})/3$ (18) 0.6931 (= ln 2)

(19) $\frac{1}{8}(\frac{1}{4})^6$ (20) 0.296   (21) $2/\sqrt{3}$

(22) 0.3837

Exercises 7.12 (Page 207)

(1) 0   (2) 293.3   (3) $\frac{8}{15}$

(4) $-\frac{4}{9}$ (5) 1.3863(= ln (4))

(6) 0   (7) $-\frac{1}{3}$ (8) $-\frac{3}{8}$

Exercises 7.13 (Page 210)

(1) 156

(2) $\frac{B}{\lambda}\{\exp[-\lambda t] - 1\} - \frac{A}{\theta}\{\exp[-\theta t] - 1\}$

Exercises 7.14 (Page 215)

(1) $2 \exp [x^{1/2}] + C$

(2) $-(1 - x^2)^{1/2} + C$

(3) $(1 + e^x)[\ln (1 + e^x) - 1] + C$

(4) $\frac{2}{3}(x + 3)^{3/2} + C$

(5) $\frac{1}{3}(1 + x^2)^{3/2} + C$

(6) $\ln (1 + e^x) + C$

(7) $\dfrac{\sin^4 x}{4} + C$

(8) $-\ln (\cos x) + C$

(9) $\frac{1}{4} \ln (x^4 - 1) + C$

(10) $\frac{1}{8}(a^2 + x^2)^4 + C$

(11) $-\dfrac{\cos^4 x}{4} + C$

(12) $\dfrac{1}{4(a^2 - x^2)^2}$

(13) $-\dfrac{1}{12}(a^3 - x^3)^4 + C$

(14) $\dfrac{1}{4a^2} \ln \left\{ \dfrac{a^2 + x^2}{a^2 - x^2} \right\} + C$

(15) $\frac{1}{2}[\ln x]^2 + C$

(16) $\frac{1}{4}(1 + \ln x)^4 + C$

(17) $(4 \cos x - 2)^{-1} + C$

(18) $\dfrac{[\ln x]^{n+1}}{n + 1} + C$

(19) $-\dfrac{(1 - e^x)^4}{4} + C$

(20) $-(1 + \ln x)^{-1} + C$

(21) $\dfrac{1}{8a^4} \ln \left\{ \dfrac{a^4 + x^4}{a^4 - x^4} \right\} + C$

(22) $-\frac{1}{2} \exp[-x^2] + C$

Exercises 7.15 (Page 219)

(1) $\frac{1}{5} x^5 (\ln(x) - \frac{1}{5}) + C$

(2) $\dfrac{x^{n+1}}{n+1} \ln(x) - \dfrac{x^{n+1}}{(n+1)^2} + C$

(3) $-(x \cos ax)/a + (\sin ax)/a^2 + C$

(4) $x \sin(x) + \cos(x) + C$

(5) $(x^3 - 3x^2 + 6x - 6)e^x + C$

(6) $(ax - 1) \exp[ax]/a^2 + C$

(7) $-\dfrac{1}{a^3} (a^2 x^2 + 2ax + 2) \exp[-ax] + C$

(8) $-x^2 \cos(x) + 2x \sin(x) + 2 \cos(x) + C$

(9) 154.9

(10) a^{-2}

(11) 0.564 (= 6 − 2e)

(12) 0.264 (= 1 − 2/e)

Exercises 7.16 (Page 222)

(1) $2 \ln(x + 1) + 4 \ln(x - 2) + C.$

(2) $x + \dfrac{3}{2} \ln \left\{ \dfrac{x-1}{x+1} \right\} + C$

(3) $-\frac{1}{3}\ln(x) - \frac{7}{6}\ln(x+3) + \frac{3}{2}\ln(x-1) + C$

(4) $\frac{1}{5}\ln\left\{\frac{x-2}{x+3}\right\} + C.$

(5) $\frac{1}{9}\ln\frac{x-4}{x+5} + C$

(6) $\frac{1}{2}\ln(x^2 - 1) + C$

(7) $x + \ln\frac{x-2}{x+2} + C$

(8) $\frac{1}{a-b}\ln\frac{x-a}{x-b} + C$

(9) $\ln\frac{x-2}{x-3} + C$

(10) $5\ln(x-2) - 12(x-2)^{-1} + C.$

(11) $\frac{1}{8}\ln\frac{x-3}{3x-1} + C.$

(12) $\frac{1}{14}\ln\frac{x-7}{x+7} + C$

(13) $\frac{3}{11}\ln\frac{x-6}{x+5} + C$

(14) $\frac{1}{5}\ln\frac{x-1}{3x+2} + C$

(15) $\frac{x^2}{2} - x + \frac{64}{9}\ln(x-4) + \frac{125}{9}\ln(x+5) + C$

(16) $\frac{1}{5}\ln\frac{x}{5-2x} + C$

Exercises 7.17 (Page 227)

(1) $\frac{1}{2}$

(2) $\sin (x) [\ln (\sin x) - 1] + C$

(3) 0.5954

Exercises 7.18 (Page 231)

(1) The exact answer is 19.5. The nearness to this value of the answers obtained by numerical integration will depend upon the number of strips into which the area is divided.

(2) 155.5

(3) The exact answer is $1 - 1/e$. Which to three figures is 0.632

Exercises 8.2 (Page 236)

(1) $y = C \exp [x] + K$

(2) $y = \frac{1}{3} \ln (3x + C)$

(3) $y = \dfrac{C \exp [bx] - a}{b^2} - \dfrac{ax}{b}$

Exercises 8.7 (Page 248)

(1) $y^2 + 2ay = x^2 + 2ax + C$

(2) $y^2 = Cx^2 - 1$

(3) $y = C \exp [\frac{1}{2} x^2] - b$

(4) Put $z = y/x$

so that $\dfrac{dz}{dx} = \dfrac{1}{x} \cdot \dfrac{dy}{dx} - \dfrac{y}{x^2}$

We arrange the given equation to the form

$$\frac{dy}{dx} = y + \frac{y}{x}$$

and we obtain

$$\frac{dz}{dx} = \frac{1}{x}\left[y + \frac{y}{x}\right] - \frac{y}{x^2}$$

$$= \frac{y}{x} = z.$$

This simple differential equation in z is solved and the solution takes the form

$$\ln(z) = x + C$$

or

$$\ln(y/x) = x + C.$$

We take exponentials of both sides and obtain

$$\frac{y}{x} = C \exp[x]$$

or, finally,

$$y = Cx \exp[x].$$

Exercises 8.8 (Page 254)

(1) $y^2 + 2xy - x^2 - 2x + 2y + C = 0$

(2) $3 \ln(2x + 2y + 3) = 2(y - x) + C$

(3) $x^2 + xy - y^2 + C = 0$

(4) $y^2 - 4y + 2xy - 4x - x^2 + C = 0$

(5) $\quad K\ln\left\{\dfrac{y + x[K - (K^2 - 1)^{1/2}]}{y + x[K + (K^2 - 1)^{1/2}]}\right\} + 2(K^2 - 1)^{1/2}\ln(y^2 + 2Kxy + x^2) + C$

Exercises 8.9 (Page 256)

(1) $y^2 = 2x^2(\ln x + C)$

(2) $y = \dfrac{x^2}{3} + \dfrac{C}{x}$

(3) $y = \dfrac{x}{1 + Cx^2}$

(4) $y = \dfrac{x^3}{4} + \dfrac{A}{x}$

(5) $y = C(1 + x^2)^{-1/2} + 1$

(6) $y = \frac{1}{2} x + \dfrac{C}{x}$

Exercises 8.10 (Page 257)

(1) $y = \dfrac{x^{n+2}}{(n+2)(n+1)} + Ax + B$

(2) $y = A[x \ln(x) - x] + Bx + C$

(3) $y = \exp[x] + Ax + B$

(4) $y = (x - 2)e^x + Ax + B$

(5) $y = -\ln(x) + Ax + B$

(6) $y = x^2 \left(\dfrac{\ln(x)}{2} - \dfrac{3}{4} \right) + Ax + B$

APPENDIX

SOME USEFUL DATA

1 cm = 0.39370 in.
1 in = 2.539998 cm.
1 in (United States) = 2.540005 cm.
1 km = 0.62137 mile.
1 mile = 1.60934 km.
1 mile (U.S.) = 1.60935 km.
1 gal = 277.3 cu in = 4.54596 litre.
1 gal (U.S.) = 231.0 cu in = 3.7853 litre.
1 lb = 453.5924 g.
π = 3.14159.
e = 2.71828.
π^2 = 9.86960.

RADIOISOTOPE HALF-LIVES

Element	Isotope	Half-life
Calcium	45	164 days
Carbon	14	5760 years
Chlorine	36	310,000 years
Chromium	51	27.8 days
Cobalt	60	5.25 years
Iodine	131	8 days
Iron	55	2.9 years
Iron	59	45 days
Phosphorus	32	14.2 days
Potassium	42	12.45 hours
Sodium	22	2.6 years
Sodium	24	15 hours
Tritium	3	12.26 years
Zinc	65	245 days

All Tables on following pages are reproduced from 'Logarithmic and Other Tables for Schools' by kind permission of the Author, Frank Castle, M.I.Mech.E., and the publishers Macmillan & Co. Ltd.

COMMON LOGARITHMS

Mean Differences

	0	1	2	3	4	5	6	7	8	9	1 2 3	4 5 6	7 8 9
10	0000	0043	0086	0128	0170	0212	0253	0294	0334	0374	5 9 13	17 21 26	30 34 38
											4 8 12	16 20 24	28 32 36
11	0414	0453	0492	0531	0569	0607	0645	0682	0719	0755	4 8 12	16 20 23	27 31 35
											4 7 11	15 18 22	26 29 33
12	0792	0828	0864	0899	0934	0969	1004	1038	1072	1106	3 7 11	14 18 21	25 28 32
											3 7 10	14 17 20	24 27 31
13	1139	1173	1206	1239	1271	1303	1335	1367	1399	1430	3 6 10	13 16 19	23 26 29
											3 7 10	13 16 19	22 25 29
14	1461	1492	1523	1553	1584	1614	1644	1673	1703	1732	3 6 9	12 15 19	22 25 28
											3 6 9	12 14 17	20 23 26
15	1761	1790	1818	1847	1875	1903	1931	1959	1987	2014	3 6 9	11 14 17	20 23 26
											3 6 8	11 14 17	19 22 25
16	2041	2068	2095	2122	2148	2175	2201	2227	2253	2279	3 6 8	11 14 16	19 22 24
											3 5 8	10 13 16	18 21 23
17	2304	2330	2355	2380	2405	2430	2455	2480	2504	2529	3 5 8	10 13 15	18 20 23
											3 5 8	10 12 15	17 20 22
18	2553	2577	2601	2625	2648	2672	2695	2718	2742	2765	2 5 7	9 12 14	17 19 21
											2 4 7	9 11 14	16 18 21
19	2788	2810	2833	2856	2878	2900	2923	2945	2967	2989	2 4 7	9 11 13	16 18 20
											2 4 6	8 11 13	15 17 19
20	3010	3032	3054	3075	3096	3118	3139	3160	3181	3201	2 4 6	8 11 13	15 17 19
21	3222	3243	3263	3284	3304	3324	3345	3365	3385	3404	2 4 6	8 10 12	14 16 18
22	3424	3444	3464	3483	3502	3522	3541	3560	3579	3598	2 4 6	8 10 12	14 15 17
23	3617	3636	3655	3674	3692	3711	3729	3747	3766	3784	2 4 6	7 9 11	13 15 17
24	3802	3820	3838	3856	3874	3892	3909	3927	3945	3962	2 4 5	7 9 11	12 14 16
25	3979	3997	4014	4031	4048	4065	4082	4099	4116	4133	2 3 5	7 9 10	12 14 15
26	4150	4166	4183	4200	4216	4232	4249	4265	4281	4298	2 3 5	7 8 10	11 13 15
27	4314	4330	4346	4362	4378	4393	4409	4425	4440	4456	2 3 5	6 8 9	11 13 14
28	4472	4487	4502	4518	4533	4548	4564	4579	4594	4609	2 3 5	6 8 9	11 12 14
29	4624	4639	4654	4669	4683	4698	4713	4728	4742	4757	1 3 4	6 7 9	10 12 13
30	4771	4786	4800	4814	4829	4843	4857	4871	4886	4900	1 3 4	6 7 9	10 11 13
31	4914	4928	4942	4955	4969	4983	4997	5011	5024	5038	1 3 4	6 7 8	10 11 12
32	5051	5065	5079	5092	5105	5119	5132	5145	5159	5172	1 3 4	5 7 8	9 11 12
33	5185	5198	5211	5224	5237	5250	5263	5276	5289	5302	1 3 4	5 6 8	9 10 12
34	5315	5328	5340	5353	5366	5378	5391	5403	5416	5428	1 3 4	5 6 8	9 10 11
35	5441	5453	5465	5478	5490	5502	5514	5527	5539	5551	1 2 4	5 6 7	9 10 11
36	5563	5575	5587	5599	5611	5623	5635	5647	5658	5670	1 2 4	5 6 7	8 10 11
37	5682	5694	5705	5717	5729	5740	5752	5763	5775	5786	1 2 3	5 6 7	8 9 10
38	5798	5809	5821	5832	5843	5855	5866	5877	5888	5899	1 2 3	5 6 7	8 9 10
39	5911	5922	5933	5944	5955	5966	5977	5988	5999	6010	1 2 3	4 5 7	8 9 10
40	6021	6031	6042	6053	6064	6075	6085	6096	6107	6117	1 2 3	4 5 6	8 9 10
41	6128	6138	6149	6160	6170	6180	6191	6201	6212	6222	1 2 3	4 5 6	7 8 9
42	6232	6243	6253	6263	6274	6284	6294	6304	6314	6325	1 2 3	4 5 6	7 8 9
43	6335	6345	6355	6365	6375	6385	6395	6405	6415	6425	1 2 3	4 5 6	7 8 9
44	6435	6444	6454	6464	6474	6484	6493	6503	6513	6522	1 2 3	4 5 6	7 8 9
45	6532	6542	6551	6561	6571	6580	6590	6599	6609	6618	1 2 3	4 5 6	7 8 9
46	6628	6637	6646	6656	6665	6675	6684	6693	6702	6712	1 2 3	4 5 6	7 7 8
47	6721	6730	6739	6749	6758	6767	6776	6785	6794	6803	1 2 3	4 5 5	6 7 8
48	6812	6821	6830	6839	6848	6857	6866	6875	6884	6893	1 2 3	4 4 5	6 7 8
49	6902	6911	6920	6928	6937	6946	6955	6964	6972	6981	1 2 3	4 4 5	6 7 8

COMMON LOGARITHMS

Mean Differences

	0	1	2	3	4	5	6	7	8	9	1 2 3	4 5 6	7 8 9
50	6990	6998	7007	7016	7024	7033	7042	7050	7059	7067	1 2 3	3 4 5	6 7 8
51	7076	7084	7093	7101	7110	7118	7126	7135	7143	7152	1 2 3	3 4 5	6 7 8
52	7160	7168	7177	7185	7193	7202	7210	7218	7226	7235	1 2 2	3 4 5	6 7 7
53	7243	7251	7259	7267	7275	7284	7292	7300	7308	7316	1 2 2	3 4 5	6 6 7
54	7324	7332	7340	7348	7356	7364	7372	7380	7388	7396	1 2 2	3 4 5	6 6 7
55	7404	7412	7419	7427	7435	7443	7451	7459	7466	7474	1 2 2	3 4 5	5 6 7
56	7482	7490	7497	7505	7513	7520	7528	7536	7543	7551	1 2 2	3 4 5	5 6 7
57	7559	7566	7574	7582	7589	7597	7604	7612	7619	7627	1 2 2	3 4 5	5 6 7
58	7634	7642	7649	7657	7664	7672	7679	7686	7694	7701	1 1 2	3 4 4	5 6 7
59	7709	7716	7723	7731	7738	7745	7752	7760	7767	7774	1 1 2	3 4 4	5 6 7
60	7782	7789	7796	7803	7810	7818	7825	7832	7839	7846	1 1 2	3 4 4	5 6 6
61	7853	7860	7868	7875	7882	7889	7896	7903	7910	7917	1 1 2	3 4 4	5 6 6
62	7924	7931	7938	7945	7952	7959	7966	7973	7980	7987	1 1 2	3 3 4	5 6 6
63	7993	8000	8007	8014	8021	8028	8035	8041	8048	8055	1 1 2	3 3 4	5 5 6
64	8062	8069	8075	8082	8089	8096	8102	8109	8116	8122	1 1 2	3 3 4	5 5 6
65	8129	8136	8142	8149	8156	8162	8169	8176	8182	8189	1 1 2	3 3 4	5 5 6
66	8195	8202	8209	8215	8222	8228	8235	8241	8248	8254	1 1 2	3 3 4	5 5 6
67	8261	8267	8274	8280	8287	8293	8299	8306	8312	8319	1 1 2	3 3 4	5 5 6
68	8325	8331	8338	8344	8351	8357	8363	8370	8376	8382	1 1 2	3 3 4	4 5 6
69	8388	8395	8401	8407	8414	8420	8426	8432	8439	8445	1 1 2	2 3 4	4 5 6
70	8451	8457	8463	8470	8476	8482	8488	8494	8500	8506	1 1 2	2 3 4	4 5 6
71	8513	8519	8525	8531	8537	8543	8549	8555	8561	8567	1 1 2	2 3 4	4 5 5
72	8573	8579	8585	8591	8597	8603	8609	8615	8621	8627	1 1 2	2 3 4	4 5 5
73	8633	8639	8645	8651	8657	8663	8669	8675	8681	8686	1 1 2	2 3 4	4 5 5
74	8692	8698	8704	8710	8716	8722	8727	8733	8739	8745	1 1 2	2 3 4	4 5 5
75	8751	8756	8762	8768	8774	8779	8785	8791	8797	8802	1 1 2	2 3 3	4 5 5
76	8808	8814	8820	8825	8831	8837	8842	8848	8854	8859	1 1 2	2 3 3	4 5 5
77	8865	8871	8876	8882	8887	8893	8899	8904	8910	8915	1 1 2	2 3 3	4 4 5
78	8921	8927	8932	8938	8943	8949	8954	8960	8965	8971	1 1 2	2 3 3	4 4 5
79	8976	8982	8987	8993	8998	9004	9009	9015	9020	9025	1 1 2	2 3 3	4 4 5
80	9031	9036	9042	9047	9053	9058	9063	9069	9074	9079	1 1 2	2 3 3	4 4 5
81	9085	9090	9096	9101	9106	9112	9117	9122	9128	9133	1 1 2	2 3 3	4 4 5
82	9138	9143	9149	9154	9159	9165	9170	9175	9180	9186	1 1 2	2 3 3	4 4 5
83	9191	9196	9201	9206	9212	9217	9222	9227	9232	9238	1 1 2	2 3 3	4 4 5
84	9243	9248	9253	9258	9263	9269	9274	9279	9284	9289	1 1 2	2 3 3	4 4 5
85	9294	9299	9304	9309	9315	9320	9325	9330	9335	9340	1 1 2	2 3 3	4 4 5
86	9345	9350	9355	9360	9365	9370	9375	9380	9385	9390	1 1 2	2 3 3	4 4 5
87	9395	9400	9405	9410	9415	9420	9425	9430	9435	9440	0 1 1	2 2 3	3 4 4
88	9445	9450	9455	9460	9465	9469	9474	9479	9484	9489	0 1 1	2 2 3	3 4 4
89	9494	9499	9504	9509	9513	9518	9523	9528	9533	9538	0 1 1	2 2 3	3 4 4
90	9542	9547	9552	9557	9562	9566	9571	9576	9581	9586	0 1 1	2 2 3	3 4 4
91	9590	9595	9600	9605	9609	9614	9619	9624	9628	9633	0 1 1	2 2 3	3 4 4
92	9638	9643	9647	9652	9657	9661	9666	9671	9675	9680	0 1 1	2 2 3	3 4 4
93	9685	9689	9694	9699	9703	9708	9713	9717	9722	9727	0 1 1	2 2 3	3 4 4
94	9731	9736	9741	9745	9750	9754	9759	9763	9768	9773	0 1 1	2 2 3	3 4 4
95	9777	9782	9786	9791	9795	9800	9805	9809	9814	9818	0 1 1	2 2 3	3 4 4
96	9823	9827	9832	9836	9841	9845	9850	9854	9859	9863	0 1 1	2 2 3	3 4 4
97	9868	9872	9877	9881	9886	9890	9894	9899	9903	9908	0 1 1	2 2 3	3 4 4
98	9912	9917	9921	9926	9930	9934	9939	9943	9948	9952	0 1 1	2 2 3	3 4 4
99	9956	9961	9965	9969	9974	9978	9983	9987	9991	9996	0 1 1	2 2 3	3 3 4

NATURAL LOGARITHMS

	0	1	2	3	4	5	6	7	8	9	Mean Differences								
											1	2	3	4	5	6	7	8	9
1·0	0·0000	0099	0198	0296	0392	0488	0583	0677	0770	0862	10	19	29	38	48	57	67	76	86
1·1	·0953	1044	1133	1222	1310	1398	1484	1570	1655	1740	9	17	26	35	44	52	61	70	78
1·2	·1823	1906	1989	2070	2151	2231	2311	2390	2469	2546	8	16	24	32	40	48	56	64	72
1·3	·2624	2700	2776	2852	2927	3001	3075	3148	3221	3293	7	15	22	30	37	44	52	59	67
1·4	·3365	3436	3507	3577	3646	3716	3784	3853	3920	3988	7	14	21	28	35	41	48	55	62
1·5	·4055	4121	4187	4253	4318	4383	4447	4511	4574	4637	6	13	19	26	32	39	45	52	58
1·6	·4700	4762	4824	4886	4947	5008	5068	5128	5188	5247	6	12	18	24	30	36	42	48	55
1·7	·5306	5365	5423	5481	5539	5596	5653	5710	5766	5822	6	11	17	24	29	34	40	46	51
1·8	·5878	5933	5988	6043	6098	6152	6206	6259	6313	6366	5	11	16	22	27	32	38	43	49
1·9	·6419	6471	6523	6575	6627	6678	6729	6780	6831	6881	5	10	15	20	26	31	36	41	46
2·0	·6931	6981	7031	7080	7129	7178	7227	7275	7324	7372	5	10	15	20	24	29	34	39	44
2·1	·7419	7467	7514	7561	7608	7655	7701	7747	7793	7839	5	9	14	19	23	28	33	37	42
2·2	·7885	7930	7975	8020	8065	8109	8154	8198	8242	8286	4	9	13	18	22	27	31	36	40
2·3	·8329	8372	8416	8459	8502	8544	8587	8629	8671	8713	4	9	13	17	21	26	30	34	38
2·4	·8755	8796	8838	8879	8920	8961	9002	9042	9083	9123	4	8	12	16	20	24	29	33	37
2·5	·9163	9203	9243	9282	9322	9361	9400	9439	9478	9517	4	8	12	16	20	24	27	31	35
2·6	·9555	9594	9632	9670	9708	9746	9783	9821	9858	9895	4	8	11	15	19	23	26	30	34
2·7	·9933	9969	1·0006	0043	0080	0116	0152	0188	0225	0260	4	7	11	15	18	22	25	29	33
2·8	1·0296	0332	0367	0403	0438	0473	0508	0543	0578	0613	4	7	11	14	18	21	25	28	32
2·9	1·0647	0682	0716	0750	0784	0818	0852	0886	0919	0953	3	7	10	14	17	20	24	27	31
3·0	1·0986	1019	1053	1086	1119	1151	1184	1217	1249	1282	3	7	10	13	16	20	23	26	30
3·1	1·1314	1346	1378	1410	1442	1474	1506	1537	1569	1600	3	6	10	13	16	19	22	25	29
3·2	1·1632	1663	1694	1725	1756	1787	1817	1848	1878	1909	3	6	9	12	15	18	22	25	28
3·3	1·1939	1969	1·2000	2030	2060	2090	2119	2149	2179	2208	3	6	9	12	15	18	21	24	27
3·4	1·2238	2267	2296	2326	2355	2384	2413	2442	2470	2499	3	6	9	12	15	17	20	23	26
3·5	1·2528	2556	2585	2613	2641	2669	2698	2726	2754	2782	3	6	8	11	14	17	20	23	25
3·6	1·2809	2837	2865	2892	2920	2947	2975	3002	3029	3056	3	5	8	11	14	16	19	22	25
3·7	1·3083	3110	3137	3164	3191	3218	3244	3271	3297	3324	3	5	8	11	13	16	19	21	24
3·8	1·3350	3376	3403	3429	3455	3481	3507	3533	3558	3584	3	5	8	10	13	16	18	21	23
3·9	1·3610	3635	3661	3686	3712	3737	3762	3788	3813	3838	3	5	8	10	13	15	18	20	23
4·0	1·3863	3888	3913	3938	3962	3987	4012	4036	4061	4085	2	5	7	10	12	15	17	20	22
4·1	1·4110	4134	4159	4183	4207	4231	4255	4279	4303	4327	2	5	7	10	12	14	17	19	22
4·2	1·4351	4375	4398	4422	4446	4469	4493	4516	4540	4563	2	5	7	9	12	14	16	19	21
4·3	1·4586	4609	4633	4656	4679	4702	4725	4748	4770	4793	2	5	7	9	12	14	16	18	21
4·4	1·4816	4839	4861	4884	4907	4929	4951	4974	4996	5019	2	5	7	9	11	14	16	18	20
4·5	1·5041	5063	5085	5107	5129	5151	5173	5195	5217	5239	2	4	7	9	11	13	15	18	20
4·6	1·5261	5282	5304	5326	5347	5369	5390	5412	5433	5454	2	4	6	9	11	13	15	17	19
4·7	1·5476	5497	5518	5539	5560	5581	5602	5623	5644	5665	2	4	6	8	11	13	15	17	19
4·8	1·5686	5707	5728	5748	5769	5790	5810	5831	5851	5872	2	4	6	8	10	12	14	16	19
4·9	1·5892	5913	5933	5953	5974	5994	6014	6034	6054	6074	2	4	6	8	10	12	14	16	18
5·0	1·6094	6114	6134	6154	6174	6194	6214	6233	6253	6273	2	4	6	8	10	12	14	16	18
5·1	1·6292	6312	6332	6351	6371	6390	6409	6429	6448	6467	2	4	6	8	10	12	14	16	18
5·2	1·6487	6506	6525	6544	6563	6582	6601	6620	6639	6658	2	4	6	8	10	11	13	15	17
5·3	1·6677	6696	6715	6734	6752	6771	6790	6808	6827	6845	2	4	6	7	9	11	13	15	17
5·4	1·6864	6882	6901	6919	6938	6956	6974	6993	7011	7029	2	4	5	7	9	11	13	15	17

Logarithms of 10^{+n}.

n	1	2	3	4	5	6	7	8	9
$\log_e 10^n$	2·3026	4·6052	6·9078	9·2103	11·5129	13·8155	16·1181	18·4207	20·7233

NATURAL LOGARITHMS

	0	1	2	3	4	5	6	7	8	9	1 2 3	4 5 6	7 8 9
5.5	1·7047	7066	7084	7102	7120	7138	7156	7174	7192	7210	2 4 5	7 9 11	13 14 16
5.6	1·7228	7246	7263	7281	7299	7317	7334	7352	7370	7387	2 4 5	7 9 11	12 14 16
5.7	1·7405	7422	7440	7457	7475	7492	7509	7527	7544	7561	2 3 5	7 9 10	12 14 16
5.8	1·7579	7596	7613	7630	7647	7664	7681	7699	7716	7733	2 3 5	7 9 10	12 14 15
5.9	1·7750	7766	7783	7800	7817	7834	7851	7867	7884	7901	2 3 5	7 8 10	12 13 15
6.0	1·7918	7934	7951	7967	7984	8001	8017	8034	8050	8066	2 3 5	7 8 10	12 13 15
6.1	1·8083	8099	8116	8132	8148	8165	8181	8197	8213	8229	2 3 5	6 8 10	11 13 15
6.2	1·8245	8262	8278	8294	8310	8326	8342	8358	8374	8390	2 3 5	6 8 10	11 13 14
6.3	1·8405	8421	8437	8453	8469	8485	8500	8516	8532	8547	2 3 5	6 8 9	11 13 14
6.4	1·8563	8579	8594	8610	8625	8641	8656	8672	8687	8703	2 3 5	6 8 9	11 12 14
6.5	1·8718	8733	8749	8764	8779	8795	8810	8825	8840	8856	2 3 5	6 8 9	11 12 14
6.6	1·8871	8886	8901	8916	8931	8946	8961	8976	8991	9006	2 3 5	6 8 9	11 12 14
6.7	1·9021	9036	9051	9066	9081	9095	9110	9125	9140	9155	1 3 4	6 7 9	10 12 13
6.8	1·9169	9184	9199	9213	9228	9242	9257	9272	9286	9301	1 3 4	6 7 9	10 12 13
6.9	1·9315	9330	9344	9359	9373	9387	9402	9416	9430	9445	1 3 4	6 7 9	10 12 13
7.0	1·9459	9473	9488	9502	9516	9530	9544	9559	9573	9587	1 3 4	6 7 9	10 11 13
7.1	1·9601	9615	9629	9643	9657	9671	9685	9699	9713	9727	1 3 4	6 7 8	10 11 13
7.2	1·9741	9755	9769	9782	9796	9810	9824	9838	9851	9865	1 3 4	6 7 8	10 11 12
7.3	1·9879	9892	9906	9920	9933	9947	9961	9974	9988	2·0001	1 3 4	5 7 8	10 11 12
7.4	2·0015	0028	0042	0055	0069	0082	0096	0109	0122	0136	1 3 4	5 7 8	9 11 12
7.5	2·0149	0162	0176	0189	0202	0215	0229	0242	0255	0268	1 3 4	5 7 8	9 11 12
7.6	2·0281	0295	0308	0321	0334	0347	0360	0373	0386	0399	1 3 4	5 7 8	9 10 12
7.7	2·0412	0425	0438	0451	0464	0477	0490	0503	0516	0528	1 3 4	5 6 8	9 10 12
7.8	2·0541	0554	0567	0580	0592	0605	0618	0631	0643	0656	1 3 4	5 6 8	9 10 11
7.9	2·0669	0681	0694	0707	0719	0732	0744	0757	0769	0782	1 3 4	5 6 8	9 10 11
8.0	2·0794	0807	0819	0832	0844	0857	0869	0882	0894	0906	1 3 4	5 6 7	9 10 11
8.1	2·0919	0931	0943	0956	0968	0980	0992	1005	1017	1029	1 2 4	5 6 7	9 10 11
8.2	2·1041	1054	1066	1078	1090	1102	1114	1126	1138	1150	1 2 4	5 6 7	9 10 11
8.3	2·1163	1175	1187	1199	1211	1223	1235	1247	1258	1270	1 2 4	5 6 7	8 10 11
8.4	2·1282	1294	1306	1318	1330	1342	1353	1365	1377	1389	1 2 4	5 6 7	8 9 11
8.5	2·1401	1412	1424	1436	1448	1459	1471	1483	1494	1506	1 2 4	5 6 7	8 9 11
8.6	2·1518	1529	1541	1552	1564	1576	1587	1599	1610	1622	1 2 3	5 6 7	8 9 10
8.7	2·1633	1645	1656	1668	1679	1691	1702	1713	1725	1736	1 2 3	5 6 7	8 9 10
8.8	2·1748	1759	1770	1782	1793	1804	1815	1827	1838	1849	1 2 3	5 6 7	8 9 10
8.9	2·1861	1872	1883	1894	1905	1917	1928	1939	1950	1961	1 2 3	4 6 7	8 9 10
9.0	2·1972	1983	1994	2006	2017	2028	2039	2050	2061	2072	1 2 3	4 6 7	8 9 10
9.1	2·2083	2094	2105	2116	2127	2138	2148	2159	2170	2181	1 2 3	4 5 7	8 9 10
9.2	2·2192	2203	2214	2225	2235	2246	2257	2268	2279	2289	1 2 3	4 5 6	8 9 10
9.3	2·2300	2311	2322	2332	2343	2354	2364	2375	2386	2396	1 2 3	4 5 6	7 9 10
9.4	2·2407	2418	2428	2439	2450	2460	2471	2481	2492	2502	1 2 3	4 5 6	7 8 10
9.5	2·2513	2523	2534	2544	2555	2565	2576	2586	2597	2607	1 2 3	4 5 6	7 8 9
9.6	2·2618	2628	2638	2649	2659	2670	2680	2690	2701	2711	1 2 3	4 5 6	7 8 9
9.7	2·2721	2732	2742	2752	2762	2773	2783	2793	2803	2814	1 2 3	4 5 6	7 8 9
9.8	2·2824	2834	2844	2854	2865	2875	2885	2895	2905	2915	1 2 3	4 5 6	7 8 9
9.9	2·2925	2935	2946	2956	2966	2976	2986	2996	3006	3016	1 2 3	4 5 6	7 8 9
10.0	2·3026												

SQUARE ROOTS: From 1 to 10

	0	1	2	3	4	5	6	7	8	9	Mean Differences 1 2 3	4 5 6	7 8 9
1·0	1·000	1·005	1·010	1·015	1·020	1·025	1·030	1·034	1·039	1·044	0 1 1	2 2 3	3 4 4
1·1	1·049	1·054	1·058	1·063	1·068	1·072	1·077	1·082	1·086	1·091	0 1 1	2 2 3	3 4 4
1·2	1·095	1·100	1·105	1·109	1·114	1·118	1·122	1·127	1·131	1·136	0 1 1	2 2 3	3 4 4
1·3	1·140	1·145	1·149	1·153	1·158	1·162	1·166	1·170	1·175	1·179	0 1 1	2 2 3	3 3 4
1·4	1·183	1·187	1·192	1·196	1·200	1·204	1·208	1·212	1·217	1·221	0 1 1	2 2 2	3 3 4
1·5	1·225	1·229	1·233	1·237	1·241	1·245	1·249	1·253	1·257	1·261	0 1 1	2 2 2	3 3 4
1·6	1·265	1·269	1·273	1·277	1·281	1·285	1·288	1·292	1·296	1·300	0 1 1	2 2 2	3 3 3
1·7	1·304	1·308	1·311	1·315	1·319	1·323	1·327	1·330	1·334	1·338	0 1 1	2 2 2	3 3 3
1·8	1·342	1·345	1·349	1·353	1·356	1·360	1·364	1·367	1·371	1·375	0 1 1	1 2 2	3 3 3
1·9	1·378	1·382	1·386	1·389	1·393	1·396	1·400	1·404	1·407	1·411	0 1 1	1 2 2	3 3 3
2·0	1·414	1·418	1·421	1·425	1·428	1·432	1·435	1·439	1·442	1·446	0 1 1	1 2 2	2 3 3
2·1	1·449	1·453	1·456	1·459	1·463	1·466	1·470	1·473	1·476	1·480	0 1 1	1 2 2	2 3 3
2·2	1·483	1·487	1·490	1·493	1·497	1·500	1·503	1·507	1·510	1·513	0 1 1	1 2 2	2 3 3
2·3	1·517	1·520	1·523	1·526	1·530	1·533	1·536	1·539	1·543	1·546	0 1 1	1 2 2	2 3 3
2·4	1·549	1·552	1·556	1·559	1·562	1·565	1·568	1·572	1·575	1·578	0 1 1	1 2 2	2 3 3
2·5	1·581	1·584	1·587	1·591	1·594	1·597	1·600	1·603	1·606	1·609	0 1 1	1 2 2	2 3 3
2·6	1·612	1·616	1·619	1·622	1·625	1·628	1·631	1·634	1·637	1·640	0 1 1	1 2 2	2 2 3
2·7	1·643	1·646	1·649	1·652	1·655	1·658	1·661	1·664	1·667	1·670	0 1 1	1 2 2	2 2 3
2·8	1·673	1·676	1·679	1·682	1·685	1·688	1·691	1·694	1·697	1·700	0 1 1	1 1 2	2 2 3
2·9	1·703	1·706	1·709	1·712	1·715	1·718	1·720	1·723	1·726	1·729	0 1 1	1 1 2	2 2 3
3·0	1·732	1·735	1·738	1·741	1·744	1·746	1·749	1·752	1·755	1·758	0 1 1	1 1 2	2 2 3
3·1	1·761	1·764	1·766	1·769	1·772	1·775	1·778	1·780	1·783	1·786	0 1 1	1 1 2	2 2 3
3·2	1·789	1·792	1·794	1·797	1·800	1·803	1·806	1·808	1·811	1·814	0 1 1	1 1 2	2 2 2
3·3	1·817	1·819	1·822	1·825	1·828	1·830	1·833	1·836	1·838	1·841	0 1 1	1 1 2	2 2 2
3·4	1·844	1·847	1·849	1·852	1·855	1·857	1·860	1·863	1·865	1·868	0 1 1	1 1 2	2 2 2
3·5	1·871	1·873	1·876	1·879	1·881	1·884	1·887	1·889	1·892	1·895	0 1 1	1 1 2	2 2 2
3·6	1·897	1·900	1·903	1·905	1·908	1·910	1·913	1·916	1·918	1·921	0 1 1	1 1 2	2 2 2
3·7	1·924	1·926	1·929	1·931	1·934	1·936	1·939	1·942	1·944	1·947	0 1 1	1 1 2	2 2 2
3·8	1·949	1·952	1·954	1·957	1·960	1·962	1·965	1·967	1·970	1·972	0 1 1	1 1 2	2 2 2
3·9	1·975	1·977	1·980	1·982	1·985	1·987	1·990	1·992	1·995	1·997	0 1 1	1 1 2	2 2 2
4·0	2·000	2·002	2·005	2·007	2·010	2·012	2·015	2·017	2·020	2·022	0 0 1	1 1 1	2 2 2
4·1	2·025	2·027	2·030	2·032	2·035	2·037	2·040	2·042	2·045	2·047	0 0 1	1 1 1	2 2 2
4·2	2·049	2·052	2·054	2·057	2·059	2·062	2·064	2·066	2·069	2·071	0 0 1	1 1 1	2 2 2
4·3	2·074	2·076	2·078	2·081	2·083	2·086	2·088	2·090	2·093	2·095	0 0 1	1 1 1	2 2 2
4·4	2·098	2·100	2·102	2·105	2·107	2·110	2·112	2·114	2·117	2·119	0 0 1	1 1 1	2 2 2
4·5	2·121	2·124	2·126	2·128	2·131	2·133	2·135	2·138	2·140	2·142	0 0 1	1 1 1	2 2 2
4·6	2·145	2·147	2·149	2·152	2·154	2·156	2·159	2·161	2·163	2·166	0 0 1	1 1 1	2 2 2
4·7	2·168	2·170	2·173	2·175	2·177	2·179	2·182	2·184	2·186	2·189	0 0 1	1 1 1	2 2 2
4·8	2·191	2·193	2·195	2·198	2·200	2·202	2·205	2·207	2·209	2·211	0 0 1	1 1 1	2 2 2
4·9	2·214	2·216	2·218	2·220	2·223	2·225	2·227	2·229	2·232	2·234	0 0 1	1 1 1	2 2 2
5·0	2·236	2·238	2·241	2·243	2·245	2·247	2·249	2·252	2·254	2·256	0 0 1	1 1 1	2 2 2
5·1	2·258	2·261	2·263	2·265	2·267	2·269	2·272	2·274	2·276	2·278	0 0 1	1 1 1	2 2 2
5·2	2·280	2·283	2·285	2·287	2·289	2·291	2·293	2·296	2·298	2·300	0 0 1	1 1 1	2 2 2
5·3	2·302	2·304	2·307	2·309	2·311	2·313	2·315	2·317	2·319	2·322	0 0 1	1 1 1	2 2 2
5·4	2·324	2·326	2·328	2·330	2·332	2·335	2·337	2·339	2·341	2·343	0 0 1	1 1 1	1 2 2

SQUARE ROOTS: From 1 to 10

	0	1	2	3	4	5	6	7	8	9	Mean Differences		
											1 2 3	4 5 6	7 8 9
5·5	2·345	2·347	2·349	2·352	2·354	2·356	2·358	2·360	2·362	2·364	0 0 1	1 1 1	1 2 2
5·6	2·366	2·369	2·371	2·373	2·375	2·377	2·379	2·381	2·383	2·385	0 0 1	1 1 1	1 2 2
5·7	2·387	2·390	2·392	2·394	2·396	2·398	2·400	2·402	2·404	2·406	0 0 1	1 1 1	1 2 2
5·8	2·408	2·410	2·412	2·415	2·417	2·419	2·421	2·423	2·425	2·427	0 0 1	1 1 1	1 2 2
5·9	2·429	2·431	2·433	2·435	2·437	2·439	2·441	2·443	2·445	2·447	0 0 1	1 1 1	1 2 2
6·0	2·449	2·452	2·454	2·456	2·458	2·460	2·462	2·464	2·466	2·468	0 0 1	1 1 1	1 2 2
6·1	2·470	2·472	2·474	2·476	2·478	2·480	2·482	2·484	2·486	2·488	0 0 1	1 1 1	1 2 2
6·2	2·490	2·492	2·494	2·496	2·498	2·500	2·502	2·504	2·506	2·508	0 0 1	1 1 1	1 2 2
6·3	2·510	2·512	2·514	2·516	2·518	2·520	2·522	2·524	2·526	2·528	0 0 1	1 1 1	1 2 2
6·4	2·530	2·532	2·534	2·536	2·538	2·540	2·542	2·544	2·546	2·548	0 0 1	1 1 1	1 2 2
6·5	2·550	2·551	2·553	2·555	2·557	2·559	2·561	2·563	2·565	2·567	0 0 1	1 1 1	1 2 2
6·6	2·569	2·571	2·573	2·575	2·577	2·579	2·581	2·583	2·585	2·587	0 0 1	1 1 1	1 2 2
6·7	2·588	2·590	2·592	2·594	2·596	2·598	2·600	2·602	2·604	2·606	0 0 1	1 1 1	1 2 2
6·8	2·608	2·610	2·612	2·613	2·615	2·617	2·619	2·621	2·623	2·625	0 0 1	1 1 1	1 2 2
6·9	2·627	2·629	2·631	2·632	2·634	2·636	2·638	2·640	2·642	2·644	0 0 1	1 1 1	1 2 2
7·0	2·646	2·648	2·650	2·651	2·653	2·655	2·657	2·659	2·661	2·663	0 0 1	1 1 1	1 2 2
7·1	2·665	2·666	2·668	2·670	2·672	2·674	2·676	2·678	2·680	2·681	0 0 1	1 1 1	1 1 2
7·2	2·683	2·685	2·687	2·689	2·691	2·693	2·694	2·696	2·698	2·700	0 0 1	1 1 1	1 1 2
7·3	2·702	2·704	2·706	2·707	2·709	2·711	2·713	2·715	2·717	2·718	0 0 1	1 1 1	1 1 2
7·4	2·720	2·722	2·724	2·726	2·728	2·729	2·731	2·733	2·735	2·737	0 0 1	1 1 1	1 1 2
7·5	2·739	2·740	2·742	2·744	2·746	2·748	2·750	2·751	2·753	2·755	0 0 1	1 1 1	1 1 2
7·6	2·757	2·759	2·760	2·762	2·764	2·766	2·768	2·769	2·771	2·773	0 0 1	1 1 1	1 1 2
7·7	2·775	2·777	2·778	2·780	2·782	2·784	2·786	2·787	2·789	2·791	0 0 1	1 1 1	1 1 2
7·8	2·793	2·795	2·796	2·798	2·800	2·802	2·804	2·805	2·807	2·809	0 0 1	1 1 1	1 1 2
7·9	2·811	2·812	2·814	2·816	2·818	2·820	2·821	2·823	2·825	2·827	0 0 1	1 1 1	1 1 2
8·0	2·828	2·830	2·832	2·834	2·835	2·837	2·839	2·841	2·843	2·844	0 0 1	1 1 1	1 1 2
8·1	2·846	2·848	2·850	2·851	2·853	2·855	2·857	2·858	2·860	2·862	0 0 1	1 1 1	1 1 2
8·2	2·864	2·865	2·867	2·869	2·871	2·872	2·874	2·876	2·877	2·879	0 0 1	1 1 1	1 1 2
8·3	2·881	2·883	2·884	2·886	2·888	2·890	2·891	2·893	2·895	2·897	0 0 1	1 1 1	1 1 2
8·4	2·898	2·900	2·902	2·903	2·905	2·907	2·909	2·910	2·912	2·914	0 0 1	1 1 1	1 1 2
8·5	2·915	2·917	2·919	2·921	2·922	2·924	2·926	2·927	2·929	2·931	0 0 1	1 1 1	1 1 2
8·6	2·933	2·934	2·936	2·938	2·939	2·941	2·943	2·944	2·946	2·948	0 0 1	1 1 1	1 1 2
8·7	2·950	2·951	2·953	2·955	2·956	2·958	2·960	2·961	2·963	2·965	0 0 1	1 1 1	1 1 2
8·8	2·966	2·968	2·970	2·972	2·973	2·975	2·977	2·978	2·980	2·982	0 0 1	1 1 1	1 1 2
8·9	2·983	2·985	2·987	2·988	2·990	2·992	2·993	2·995	2·997	2·998	0 0 1	1 1 1	1 1 2
9·0	3·000	3·002	3·003	3·005	3·007	3·008	3·010	3·012	3·013	3·015	0 0 0	1 1 1	1 1 1
9·1	3·017	3·018	3·020	3·022	3·023	3·025	3·027	3·028	3·030	3·032	0 0 0	1 1 1	1 1 1
9·2	3·033	3·035	3·036	3·038	3·040	3·041	3·043	3·045	3·046	3·048	0 0 0	1 1 1	1 1 1
9·3	3·050	3·051	3·053	3·055	3·056	3·058	3·059	3·061	3·063	3·064	0 0 0	1 1 1	1 1 1
9·4	3·066	3·068	3·069	3·071	3·072	3·074	3·076	3·077	3·079	3·081	0 0 0	1 1 1	1 1 1
9·5	3·082	3·084	3·085	3·087	3·089	3·090	3·092	3·094	3·095	3·097	0 0 0	1 1 1	1 1 1
9·6	3·098	3·100	3·102	3·103	3·105	3·106	3·108	3·110	3·111	3·113	0 0 0	1 1 1	1 1 1
9·7	3·114	3·116	3·118	3·119	3·121	3·122	3·124	3·126	3·127	3·129	0 0 0	1 1 1	1 1 1
9·8	3·130	3·132	3·134	3·135	3·137	3·138	3·140	3·142	3·143	3·145	0 0 0	1 1 1	1 1 1
9·9	3·146	3·148	3·150	3·151	3·153	3·154	3·156	3·158	3·159	3·161	0 0 0	1 1 1	1 1 1

SQUARE ROOTS: From 10 to 100

	0	1	2	3	4	5	6	7	8	9	Mean Differences 1 2 3	4 5 6	7 8 9
10	3·162	3·178	3·194	3·209	3·225	3·240	3·256	3·271	3·286	3·302	2 3 5	6 8 9	11 12 14
11	3·317	3·332	3·347	3·362	3·376	3·391	3·406	3·421	3·435	3·450	1 3 4	6 7 9	10 12 13
12	3·464	3·479	3·493	3·507	3·521	3·536	3·550	3·564	3·578	3·592	1 3 4	6 7 8	10 11 13
13	3·606	3·619	3·633	3·647	3·661	3·674	3·688	3·701	3·715	3·728	1 3 4	5 7 8	10 11 12
14	3·742	3·755	3·768	3·782	3·795	3·808	3·821	3·834	3·847	3·860	1 3 4	5 7 8	9 11 12
15	3·873	3·886	3·899	3·912	3·924	3·937	3·950	3·962	3·975	3·987	1 3 4	5 6 8	9 10 11
16	4·000	4·012	4·025	4·037	4·050	4·062	4·074	4·087	4·099	4·111	1 2 4	5 6 7	9 10 11
17	4·123	4·135	4·147	4·159	4·171	4·183	4·195	4·207	4·219	4·231	1 2 4	5 6 7	8 10 11
18	4·243	4·254	4·266	4·278	4·290	4·301	4·313	4·324	4·336	4·347	1 2 3	5 6 7	8 9 10
19	4·359	4·370	4·382	4·393	4·405	4·416	4·427	4·438	4·450	4·461	1 2 3	5 6 7	8 9 10
20	4·472	4·483	4·494	4·506	4·517	4·528	4·539	4·550	4·561	4·572	1 2 3	4 6 7	8 9 10
21	4·583	4·593	4·604	4·615	4·626	4·637	4·648	4·658	4·669	4·680	1 2 3	4 5 6	8 9 10
22	4·690	4·701	4·712	4·722	4·733	4·743	4·754	4·764	4·775	4·785	1 2 3	4 5 6	7 8 9
23	4·796	4·806	4·817	4·827	4·837	4·848	4·858	4·868	4·879	4·889	1 2 3	4 5 6	7 8 9
24	4·899	4·909	4·919	4·930	4·940	4·950	4·960	4·970	4·980	4·990	1 2 3	4 5 6	7 8 9
25	5·000	5·010	5·020	5·030	5·040	5·050	5·060	5·070	5·079	5·089	1 2 3	4 5 6	7 8 9
26	5·099	5·109	5·119	5·128	5·138	5·148	5·158	5·167	5·177	5·187	1 2 3	4 5 6	7 8 9
27	5·196	5·206	5·215	5·225	5·235	5·244	5·254	5·263	5·273	5·282	1 2 3	4 5 6	7 8 9
28	5·292	5·301	5·310	5·320	5·329	5·339	5·348	5·357	5·367	5·376	1 2 3	4 5 6	7 7 8
29	5·385	5·394	5·404	5·413	5·422	5·431	5·441	5·450	5·459	5·468	1 2 3	4 5 5	6 7 8
30	5·477	5·486	5·495	5·505	5·514	5·523	5·532	5·541	5·550	5·559	1 2 3	4 4 5	6 7 8
31	5·568	5·577	5·586	5·595	5·604	5·612	5·621	5·630	5·639	5·648	1 2 3	3 4 5	6 7 8
32	5·657	5·666	5·675	5·683	5·692	5·701	5·710	5·718	5·727	5·736	1 2 3	3 4 5	6 7 8
33	5·745	5·753	5·762	5·771	5·779	5·788	5·797	5·805	5·814	5·822	1 2 3	3 4 5	6 7 8
34	5·831	5·840	5·848	5·857	5·865	5·874	5·882	5·891	5·899	5·908	1 2 3	3 4 5	6 7 8
35	5·916	5·925	5·933	5·941	5·950	5·958	5·967	5·975	5·983	5·992	1 2 2	3 4 5	6 7 8
36	6·000	6·008	6·017	6·025	6·033	6·042	6·050	6·058	6·066	6·075	1 2 2	3 4 5	6 7 7
37	6·083	6·091	6·099	6·107	6·116	6·124	6·132	6·140	6·148	6·156	1 2 2	3 4 5	6 7 7
38	6·164	6·173	6·181	6·189	6·197	6·205	6·213	6·221	6·229	6·237	1 2 2	3 4 5	6 6 7
39	6·245	6·253	6·261	6·269	6·277	6·285	6·293	6·301	6·309	6·317	1 2 2	3 4 5	6 6 7
40	6·325	6·332	6·340	6·348	6·356	6·364	6·372	6·380	6·387	6·395	1 2 2	3 4 5	6 6 7
41	6·403	6·411	6·419	6·427	6·434	6·442	6·450	6·458	6·465	6·473	1 2 2	3 4 5	5 6 7
42	6·481	6·488	6·496	6·504	6·512	6·519	6·527	6·535	6·542	6·550	1 2 2	3 4 5	5 6 7
43	6·557	6·565	6·573	6·580	6·588	6·595	6·603	6·611	6·618	6·626	1 2 2	3 4 5	5 6 7
44	6·633	6·641	6·648	6·656	6·663	6·671	6·678	6·686	6·693	6·701	1 2 2	3 4 5	5 6 7
45	6·708	6·716	6·723	6·731	6·738	6·745	6·753	6·760	6·768	6·775	1 1 2	3 4 4	5 6 7
46	6·782	6·790	6·797	6·804	6·812	6·819	6·826	6·834	6·841	6·848	1 1 2	3 4 4	5 6 7
47	6·856	6·863	6·870	6·877	6·885	6·892	6·899	6·907	6·914	6·921	1 1 2	3 4 4	5 6 7
48	6·928	6·935	6·943	6·950	6·957	6·964	6·971	6·979	6·986	6·993	1 1 2	3 4 4	5 6 6
49	7·000	7·007	7·014	7·021	7·029	7·036	7·043	7·050	7·057	7·064	1 1 2	3 4 4	5 6 6
50	7·071	7·078	7·085	7·092	7·099	7·106	7·113	7·120	7·127	7·134	1 1 2	3 4 4	5 6 6
51	7·141	7·148	7·155	7·162	7·169	7·176	7·183	7·190	7·197	7·204	1 1 2	3 4 4	5 6 6
52	7·211	7·218	7·225	7·232	7·239	7·246	7·253	7·259	7·266	7·273	1 1 2	3 3 4	5 6 6
53	7·280	7·287	7·294	7·301	7·308	7·314	7·321	7·328	7·335	7·342	1 1 2	3 3 4	5 5 6
54	7·348	7·355	7·362	7·369	7·376	7·382	7·389	7·396	7·403	7·409	1 1 2	3 3 4	5 5 6

SQUARE ROOTS: From 10 to 100

	0	1	2	3	4	5	6	7	8	9	Mean Differences								
											1	2	3	4	5	6	7	8	9
55	7·416	7·423	7·430	7·436	7·443	7·450	7·457	7·463	7·470	7·477	1	1	2	3	3	4	5	5	6
56	7·483	7·490	7·497	7·503	7·510	7·517	7·523	7·530	7·537	7·543	1	1	2	3	3	4	5	5	6
57	7·550	7·556	7·563	7·570	7·576	7·583	7·589	7·596	7·603	7·609	1	1	2	3	3	4	5	5	6
58	7·616	7·622	7·629	7·635	7·642	7·649	7·655	7·662	7·668	7·675	1	1	2	3	3	4	5	5	6
59	7·681	7·688	7·694	7·701	7·707	7·714	7·720	7·727	7·733	7·740	1	1	2	3	3	4	4	5	6
60	7·746	7·752	7·759	7·765	7·772	7·778	7·785	7·791	7·797	7·804	1	1	2	3	3	4	4	5	6
61	7·810	7·817	7·823	7·829	7·836	7·842	7·849	7·855	7·861	7·868	1	1	2	3	3	4	4	5	6
62	7·874	7·880	7·887	7·893	7·899	7·906	7·912	7·918	7·925	7·931	1	1	2	3	3	4	4	5	6
63	7·937	7·944	7·950	7·956	7·962	7·969	7·975	7·981	7·987	7·994	1	1	2	3	3	4	4	5	6
64	8·000	8·006	8·012	8·019	8·025	8·031	8·037	8·044	8·050	8·056	1	1	2	3	3	4	4	5	6
65	8·062	8·068	8·075	8·081	8·087	8·093	8·099	8·106	8·112	8·118	1	1	2	2	3	4	4	5	6
66	8·124	8·130	8·136	8·142	8·149	8·155	8·161	8·167	8·173	8·179	1	1	2	2	3	4	4	5	5
67	8·185	8·191	8·198	8·204	8·210	8·216	8·222	8·228	8·234	8·240	1	1	2	2	3	4	4	5	5
68	8·246	8·252	8·258	8·264	8·270	8·276	8·283	8·289	8·295	8·301	1	1	2	2	3	4	4	5	5
69	8·307	8·313	8·319	8·325	8·331	8·337	8·343	8·349	8·355	8·361	1	1	2	2	3	4	4	5	5
70	8·367	8·373	8·379	8·385	8·390	8·396	8·402	8·408	8·414	8·420	1	1	2	2	3	4	4	5	5
71	8·426	8·432	8·438	8·444	8·450	8·456	8·462	8·468	8·473	8·479	1	1	2	2	3	4	4	5	5
72	8·485	8·491	8·497	8·503	8·509	8·515	8·521	8·526	8·532	8·538	1	1	2	2	3	3	4	5	5
73	8·544	8·550	8·556	8·562	8·567	8·573	8·579	8·585	8·591	8·597	1	1	2	2	3	3	4	5	5
74	8·602	8·608	8·614	8·620	8·626	8·631	8·637	8·643	8·649	8·654	1	1	2	2	3	3	4	5	5
75	8·660	8·666	8·672	8·678	8·683	8·689	8·695	8·701	8·706	8·712	1	1	2	2	3	3	4	5	5
76	8·718	8·724	8·729	8·735	8·741	8·746	8·752	8·758	8·764	8·769	1	1	2	2	3	3	4	5	5
77	8·775	8·781	8·786	8·792	8·798	8·803	8·809	8·815	8·820	8·826	1	1	2	2	3	3	4	4	5
78	8·832	8·837	8·843	8·849	8·854	8·860	8·866	8·871	8·877	8·883	1	1	2	2	3	3	4	4	5
79	8·888	8·894	8·899	8·905	8·911	8·916	8·922	8·927	8·933	8·939	1	1	2	2	3	3	4	4	5
80	8·944	8·950	8·955	8·961	8·967	8·972	8·978	8·983	8·989	8·994	1	1	2	2	3	3	4	4	5
81	9·000	9·006	9·011	9·017	9·022	9·028	9·033	9·039	9·044	9·050	1	1	2	2	3	3	4	4	5
82	9·055	9·061	9·066	9·072	9·077	9·083	9·088	9·094	9·099	9·105	1	1	2	2	3	3	4	4	5
83	9·110	9·116	9·121	9·127	9·132	9·138	9·143	9·149	9·154	9·160	1	1	2	2	3	3	4	4	5
84	9·165	9·171	9·176	9·182	9·187	9·192	9·198	9·203	9·209	9·214	1	1	2	2	3	3	4	4	5
85	9·220	9·225	9·230	9·236	9·241	9·247	9·252	9·257	9·263	9·268	1	1	2	2	3	3	4	4	5
86	9·274	9·279	9·284	9·290	9·295	9·301	9·306	9·311	9·317	9·322	1	1	2	2	3	3	4	4	5
87	9·327	9·333	9·338	9·343	9·349	9·354	9·359	9·365	9·370	9·375	1	1	2	2	3	3	4	4	5
88	9·381	9·386	9·391	9·397	9·402	9·407	9·413	9·418	9·423	9·429	1	1	2	2	3	3	4	4	5
89	9·434	9·439	9·445	9·450	9·455	9·460	9·466	9·471	9·476	9·482	1	1	2	2	3	3	4	4	5
90	9·487	9·492	9·497	9·503	9·508	9·513	9·518	9·524	9·529	9·534	1	1	2	2	3	3	4	4	5
91	9·539	9·545	9·550	9·555	9·560	9·566	9·571	9·576	9·581	9·586	1	1	2	2	3	3	4	4	5
92	9·592	9·597	9·602	9·607	9·612	9·618	9·623	9·628	9·633	9·638	1	1	2	2	3	3	4	4	5
93	9·644	9·649	9·654	9·659	9·664	9·670	9·675	9·680	9·685	9·690	1	1	2	2	3	3	4	4	5
94	9·695	9·701	9·706	9·711	9·716	9·721	9·726	9·731	9·737	9·742	1	1	2	2	3	3	4	4	5
95	9·747	9·752	9·757	9·762	9·767	9·772	9·778	9·783	9·788	9·793	1	1	2	2	3	3	4	4	5
96	9·798	9·803	9·808	9·813	9·818	9·823	9·829	9·834	9·839	9·844	1	1	2	2	3	3	4	4	5
97	9·849	9·854	9·859	9·864	9·869	9·874	9·879	9·884	9·889	9·894	1	1	1	2	3	3	4	4	5
98	9·899	9·905	9·910	9·915	9·920	9·925	9·930	9·935	9·940	9·945	0	1	1	2	2	3	3	4	4
99	9·950	9·955	9·960	9·965	9·970	9·975	9·980	9·985	9·990	9·995	0	1	1	2	2	3	3	4	4

NATURAL SINES

Degrees	0' 0°.0	6' 0°.1	12' 0°.2	18' 0°.3	24' 0°.4	30' 0°.5	36' 0°.6	42' 0°.7	48' 0°.8	54' 0°.9	Mean Differences 1 2 3	4 5
0	·0000	0017	0035	0052	0070	0087	0105	0122	0140	0157	3 6 9	12 15
1	·0175	0192	0209	0227	0244	0262	0279	0297	0314	0332	3 6 9	12 15
2	·0349	0366	0384	0401	0419	0436	0454	0471	0488	0506	3 6 9	12 15
3	·0523	0541	0558	0576	0593	0610	0628	0645	0663	0680	3 6 9	12 15
4	·0698	0715	0732	0750	0767	0785	0802	0819	0837	0854	3 6 9	12 15
5	·0872	0889	0906	0924	0941	0958	0976	0993	1011	1028	3 6 9	12 14
6	·1045	1063	1080	1097	1115	1132	1149	1167	1184	1201	3 6 9	12 14
7	·1219	1236	1253	1271	1288	1305	1323	1340	1357	1374	3 6 9	12 14
8	·1392	1409	1426	1444	1461	1478	1495	1513	1530	1547	3 6 9	12 14
9	·1564	1582	1599	1616	1633	1650	1668	1685	1702	1719	3 6 9	12 14
10	·1736	1754	1771	1788	1805	1822	1840	1857	1874	1891	3 6 9	12 14
11	·1908	1925	1942	1959	1977	1994	2011	2028	2045	2062	3 6 9	11 14
12	·2079	2096	2113	2130	2147	2164	2181	2198	2215	2232	3 6 9	11 14
13	·2250	2267	2284	2300	2317	2334	2351	2368	2385	2402	3 6 8	11 14
14	·2419	2436	2453	2470	2487	2504	2521	2538	2554	2571	3 6 8	11 14
15	·2588	2605	2622	2639	2656	2672	2689	2706	2723	2740	3 6 8	11 14
16	·2756	2773	2790	2807	2823	2840	2857	2874	2890	2907	3 6 8	11 14
17	·2924	2940	2957	2974	2990	3007	3024	3040	3057	3074	3 6 8	11 14
18	·3090	3107	3123	3140	3156	3173	3190	3206	3223	3239	3 6 8	11 14
19	·3256	3272	3289	3305	3322	3338	3355	3371	3387	3404	3 5 8	11 14
20	·3420	3437	3453	3469	3486	3502	3518	3535	3551	3567	3 5 8	11 14
21	·3584	3600	3616	3633	3649	3665	3681	3697	3714	3730	3 5 8	11 14
22	·3746	3762	3778	3795	3811	3827	3843	3859	3875	3891	3 5 8	11 14
23	·3907	3923	3939	3955	3971	3987	4003	4019	4035	4051	3 5 8	11 14
24	·4067	4083	4099	4115	4131	4147	4163	4179	4195	4210	3 5 8	11 13
25	·4226	4242	4258	4274	4289	4305	4321	4337	4352	4368	3 5 8	11 13
26	·4384	4399	4415	4431	4446	4462	4478	4493	4509	4524	3 5 8	10 13
27	·4540	4555	4571	4586	4602	4617	4633	4648	4664	4679	3 5 8	10 13
28	·4695	4710	4726	4741	4756	4772	4787	4802	4818	4833	3 5 8	10 13
29	·4848	4863	4879	4894	4909	4924	4939	4955	4970	4985	3 5 8	10 13
30	·5000	5015	5030	5045	5060	5075	5090	5105	5120	5135	3 5 8	10 13
31	·5150	5165	5180	5195	5210	5225	5240	5255	5270	5284	2 5 7	10 12
32	·5299	5314	5329	5344	5358	5373	5388	5402	5417	5432	2 5 7	10 12
33	·5446	5461	5476	5490	5505	5519	5534	5548	5563	5577	2 5 7	10 12
34	·5592	5606	5621	5635	5650	5664	5678	5693	5707	5721	2 5 7	10 12
35	·5736	5750	5764	5779	5793	5807	5821	5835	5850	5864	2 5 7	10 12
36	·5878	5892	5906	5920	5934	5948	5962	5976	5990	6004	2 5 7	9 12
37	·6018	6032	6046	6060	6074	6088	6101	6115	6129	6143	2 5 7	9 12
38	·6157	6170	6184	6198	6211	6225	6239	6252	6266	6280	2 5 7	9 11
39	·6293	6307	6320	6334	6347	6361	6374	6388	6401	6414	2 4 7	9 11
40	·6428	6441	6455	6468	6481	6494	6508	6521	6534	6547	2 4 7	9 11
41	·6561	6574	6587	6600	6613	6626	6639	6652	6665	6678	2 4 7	9 11
42	·6691	6704	6717	6730	6743	6756	6769	6782	6794	6807	2 4 6	9 11
43	·6820	6833	6845	6858	6871	6884	6896	6909	6921	6934	2 4 6	8 11
44	·6947	6959	6972	6984	6997	7009	7022	7034	7046	7059	2 4 6	8 10

NATURAL SINES

Degrees	0' 0°.0	6' 0°.1	12' 0°.2	18' 0°.3	24' 0°.4	30' 0°.5	36' 0°.6	42' 0°.7	48' 0°.8	54' 0°.9	Mean Differences 1 2 3	4 5
45	·7071	7083	7096	7108	7120	7133	7145	7157	7169	7181	2 4 6	8 10
46	·7193	7206	7218	7230	7242	7254	7266	7278	7290	7302	2 4 6	8 10
47	·7314	7325	7337	7349	7361	7373	7385	7396	7408	7420	2 4 6	8 10
48	·7431	7443	7455	7466	7478	7490	7501	7513	7524	7536	2 4 6	8 10
49	·7547	7558	7570	7581	7593	7604	7615	7627	7638	7649	2 4 6	8 9
50	·7660	7672	7683	7694	7705	7716	7727	7738	7749	7760	2 4 6	7 9
51	·7771	7782	7793	7804	7815	7826	7837	7848	7859	7869	2 4 5	7 9
52	·7880	7891	7902	7912	7923	7934	7944	7955	7965	7976	2 4 5	7 9
53	·7986	7997	8007	8018	8028	8039	8049	8059	8070	8080	2 3 5	7 9
54	·8090	8100	8111	8121	8131	8141	8151	8161	8171	8181	2 3 5	7 8
55	·8192	8202	8211	8221	8231	8241	8251	8261	8271	8281	2 3 5	7 8
56	·8290	8300	8310	8320	8329	8339	8348	8358	8368	8377	2 3 5	6 8
57	·8387	8396	8406	8415	8425	8434	8443	8453	8462	8471	2 3 5	6 8
58	·8480	8490	8499	8508	8517	8526	8536	8545	8554	8563	2 3 5	6 8
59	·8572	8581	8590	8599	8607	8616	8625	8634	8643	8652	1 3 4	6 7
60	·8660	8669	8678	8686	8695	8704	8712	8721	8729	8738	1 3 4	6 7
61	·8746	8755	8763	8771	8780	8788	8796	8805	8813	8821	1 3 4	6 7
62	·8829	8838	8846	8854	8862	8870	8878	8886	8894	8902	1 3 4	5 7
63	·8910	8918	8926	8934	8942	8949	8957	8965	8973	8980	1 3 4	5 6
64	·8988	8996	9003	9011	9018	9026	9033	9041	9048	9056	1 3 4	5 6
65	·9063	9070	9078	9085	9092	9100	9107	9114	9121	9128	1 2 4	5 6
66	·9135	9143	9150	9157	9164	9171	9178	9184	9191	9198	1 2 3	5 6
67	·9205	9212	9219	9225	9232	9239	9245	9252	9259	9265	1 2 3	4 6
68	·9272	9278	9285	9291	9298	9304	9311	9317	9323	9330	1 2 3	4 5
69	·9336	9342	9348	9354	9361	9367	9373	9379	9385	9391	1 2 3	4 5
70	·9397	9403	9409	9415	9421	9426	9432	9438	9444	9449	1 2 3	4 5
71	·9455	9461	9466	9472	9478	9483	9489	9494	9500	9505	1 2 3	4 5
72	·9511	9516	9521	9527	9532	9537	9542	9548	9553	9558	1 2 3	3 4
73	·9563	9568	9573	9578	9583	9588	9593	9598	9603	9608	1 2 2	3 4
74	·9613	9617	9622	9627	9632	9636	9641	9646	9650	9655	1 2 2	3 4
75	·9659	9664	9668	9673	9677	9681	9686	9690	9694	9699	1 1 2	3 4
76	·9703	9707	9711	9715	9720	9724	9728	9732	9736	9740	1 1 2	3 3
77	·9744	9748	9751	9755	9759	9763	9767	9770	9774	9778	1 1 2	3 3
78	·9781	9785	9789	9792	9796	9799	9803	9806	9810	9813	1 1 2	2 3
79	·9816	9820	9823	9826	9829	9833	9836	9839	9842	9845	1 1 2	2 3
80	·9848	9851	9854	9857	9860	9863	9866	9869	9871	9874	0 1 1	2 2
81	·9877	9880	9882	9885	9888	9890	9893	9895	9898	9900	0 1 1	2 2
82	·9903	9905	9907	9910	9912	9914	9917	9919	9921	9923	0 1 1	2 2
83	·9925	9928	9930	9932	9934	9936	9938	9940	9942	9943	0 1 1	1 2
84	·9945	9947	9949	9951	9952	9954	9956	9957	9959	9960	0 1 1	1 2
85	·9962	9963	9965	9966	9968	9969	9971	9972	9973	9974	0 0 1	1 1
86	·9976	9977	9978	9979	9980	9981	9982	9983	9984	9985	0 0 1	1 1
87	·9986	9987	9988	9989	9990	9990	9991	9992	9993	9993	0 0 0	1 1
88	·9994	9995	9995	9996	9996	9997	9997	9997	9998	9998	0 0 0	0 0
89	·9998	9999	9999	9999	9999	1·000	1·000	1·000	1·000	1·000	0 0 0	0 0
90	1·000											

NATURAL COSINES

[Numbers in difference columns to be subtracted, not added.]

Degrees	0' 0°.0	6' 0°.1	12' 0°.2	18' 0°.3	24' 0°.4	30' 0°.5	36' 0°.6	42' 0°.7	48' 0°.8	54' 0°.9	Mean Differences 1 2 3		4 5	
0	1·000	1·000	1·000	1·000	1·000	1·000	·9999	9999	9999	9999	0 0 0		0 0	
1	·9998	9998	9998	9997	9997	9997	9996	9996	9995	9995	0 0 0		0 0	
2	·9994	9993	9993	9992	9991	9990	9990	9989	9988	9987	0 0 0		1 1	
3	·9986	9985	9984	9983	9982	9981	9980	9979	9978	9977	0 0 1		1 1	
4	·9976	9974	9973	9972	9971	9969	9968	9966	9965	9963	0 0 1		1 1	
5	·9962	9960	9959	9957	9956	9954	9952	9951	9949	9947	0 1 1		1 2	
6	·9945	9943	9942	9940	9938	9936	9934	9932	9930	9928	0 1 1		1 2	
7	·9925	9923	9921	9919	9917	9914	9912	9910	9907	9905	0 1 1		2 2	
8	·9903	9900	9898	9895	9893	9890	9888	9885	9882	9880	0 1 1		2 2	
9	·9877	9874	9871	9869	9866	9863	9860	9857	9854	9851	0 1 1		2 2	
10	·9848	9845	9842	9839	9836	9833	9829	9826	9823	9820	1 1 2		2 3	
11	·9816	9813	9810	9806	9803	9799	9796	9792	9789	9785	1 1 2		2 3	
12	·9781	9778	9774	9770	9767	9763	9759	9755	9751	9748	1 1 2		3 3	
13	·9744	9740	9736	9732	9728	9724	9720	9715	9711	9707	1 1 2		3 3	
14	·9703	9699	9694	9690	9686	9681	9677	9673	9668	9664	1 1 2		3 4	
15	·9659	9655	9650	9646	9641	9636	9632	9627	9622	9617	1 2 2		3 4	
16	·9613	9608	9603	9598	9593	9588	9583	9578	9573	9568	1 2 2		3 4	
17	·9563	9558	9553	9548	9542	9537	9532	9527	9521	9516	1 2 3		3 4	
18	·9511	9505	9500	9494	9489	9483	9478	9472	9466	9461	1 2 3		4 5	
19	·9455	9449	9444	9438	9432	9426	9421	9415	9409	9403	1 2 3		4 5	
20	·9397	9391	9385	9379	9373	9367	9361	9354	9348	9342	1 2 3		4 5	
21	·9336	9330	9323	9317	9311	9304	9298	9291	9285	9278	1 2 3		4 5	
22	·9272	9265	9259	9252	9245	9239	9232	9225	9219	9212	1 2 3		4 6	
23	·9205	9198	9191	9184	9178	9171	9164	9157	9150	9143	1 2 3		5 6	
24	·9135	9128	9121	9114	9107	9100	9092	9085	9078	9070	1 2 4		5 6	
25	·9063	9056	9048	9041	9033	9026	9018	9011	9003	8996	1 3 4		5 6	
26	·8988	8980	8973	8965	8957	8949	8942	8934	8926	8918	1 3 4		5 6	
27	·8910	8902	8894	8886	8878	8870	8862	8854	8846	8838	1 3 4		5 7	
28	·8829	8821	8813	8805	8796	8788	8780	8771	8763	8755	1 3 4		6 7	
29	·8746	8738	8729	8721	8712	8704	8695	8686	8678	8669	1 3 4		6 7	
30	·8660	8652	8643	8634	8625	8616	8607	8599	8590	8581	1 3 4		6 7	
31	·8572	8563	8554	8545	8536	8526	8517	8508	8499	8490	2 3 5		6 8	
32	·8480	8471	8462	8453	8443	8434	8425	8415	8406	8396	2 3 5		6 8	
33	·8387	8377	8368	8358	8348	8339	8329	8320	8310	8300	2 3 5		6 8	
34	·8290	8281	8271	8261	8251	8241	8231	8221	8211	8202	2 3 5		7 8	
35	·8192	8181	8171	8161	8151	8141	8131	8121	8111	8100	2 3 5		7 8	
36	·8090	8080	8070	8059	8049	8039	8028	8018	8007	7997	2 3 5		7 9	
37	·7986	7976	7965	7955	7944	7934	7923	7912	7902	7891	2 4 5		7 9	
38	·7880	7869	7859	7848	7837	7826	7815	7804	7793	7782	2 4 5		7 9	
39	·7771	7760	7749	7738	7727	7716	7705	7694	7683	7672	2 4 6		7 9	
40	·7660	7649	7638	7627	7615	7604	7593	7581	7570	7559	2 4 6		8 9	
41	·7547	7536	7524	7513	7501	7490	7478	7466	7455	7443	2 4 6		8 10	
42	·7431	7420	7408	7396	7385	7373	7361	7349	7337	7325	2 4 6		8 10	
43	·7314	7302	7290	7278	7266	7254	7242	7230	7218	7206	2 4 6		8 10	
44	·7193	7181	7169	7157	7145	7133	7120	7108	7096	7083	2 4 6		8 10	

NATURAL COSINES

[Numbers in difference columns to be subtracted, not added.]

Degrees	0′ 0°·0	6′ 0°·1	12′ 0°·2	18′ 0°·3	24′ 0°·4	30′ 0°·5	36′ 0°·6	42′ 0°·7	48′ 0°·8	54′ 0°·9	Mean Differences 1	2	3	4	5
45	·7071	7059	7046	7034	7022	7009	6997	6984	6972	6959	2	4	6	8	10
46	·6947	6934	6921	6909	6896	6884	6871	6858	6845	6833	2	4	6	8	11
47	·6820	6807	6794	6782	6769	6756	6743	6730	6717	6704	2	4	6	9	11
48	·6691	6678	6665	6652	6639	6626	6613	6600	6587	6574	2	4	7	9	11
49	·6561	6547	6534	6521	6508	6494	6481	6468	6455	6441	2	4	7	9	11
50	·6428	6414	6401	6388	6374	6361	6347	6334	6320	6307	2	4	7	9	11
51	·6293	6280	6266	6252	6239	6225	6211	6198	6184	6170	2	5	7	9	11
52	·6157	6143	6129	6115	6101	6088	6074	6060	6046	6032	2	5	7	9	12
53	·6018	6004	5990	5976	5962	5948	5934	5920	5906	5892	2	5	7	9	12
54	·5878	5864	5850	5835	5821	5807	5793	5779	5764	5750	2	5	7	9	12
55	·5736	5721	5707	5693	5678	5664	5650	5635	5621	5606	2	5	7	10	12
56	·5592	5577	5563	5548	5534	5519	5505	5490	5476	5461	2	5	7	10	12
57	·5446	5432	5417	5402	5388	5373	5358	5344	5329	5314	2	5	7	10	12
58	·5299	5284	5270	5255	5240	5225	5210	5195	5180	5165	2	5	7	10	12
59	·5150	5135	5120	5105	5090	5075	5060	5045	5030	5015	3	5	8	10	13
60	·5000	4985	4970	4955	4939	4924	4909	4894	4879	4863	3	5	8	10	13
61	·4848	4833	4818	4802	4787	4772	4756	4741	4726	4710	3	5	8	10	13
62	·4695	4679	4664	4648	4633	4617	4602	4586	4571	4555	3	5	8	10	13
63	·4540	4524	4509	4493	4478	4462	4446	4431	4415	4399	3	5	8	10	13
64	·4384	4368	4352	4337	4321	4305	4289	4274	4258	4242	3	5	8	11	13
65	·4226	4210	4195	4179	4163	4147	4131	4115	4099	4083	3	5	8	11	13
66	·4067	4051	4035	4019	4003	3987	3971	3955	3939	3923	3	5	8	11	14
67	·3907	3891	3875	3859	3843	3827	3811	3795	3778	3762	3	5	8	11	14
68	·3746	3730	3714	3697	3681	3665	3649	3633	3616	3600	3	5	8	11	14
69	·3584	3567	3551	3535	3518	3502	3486	3469	3453	3437	3	5	8	11	14
70	·3420	3404	3387	3371	3355	3338	3322	3305	3289	3272	3	5	8	11	14
71	·3256	3239	3223	3206	3190	3173	3156	3140	3123	3107	3	6	8	11	14
72	·3090	3074	3057	3040	3024	3007	2990	2974	2957	2940	3	6	8	11	14
73	·2924	2907	2890	2874	2857	2840	2823	2807	2790	2773	3	6	8	11	14
74	·2756	2740	2723	2706	2689	2672	2656	2639	2622	2605	3	6	8	11	14
75	·2588	2571	2554	2538	2521	2504	2487	2470	2453	2436	3	6	8	11	14
76	·2419	2402	2385	2368	2351	2334	2317	2300	2284	2267	3	6	8	11	14
77	·2250	2233	2215	2198	2181	2164	2147	2130	2113	2096	3	6	9	11	14
78	·2079	2062	2045	2028	2011	1994	1977	1959	1942	1925	3	6	9	11	14
79	·1908	1891	1874	1857	1840	1822	1805	1788	1771	1754	3	6	9	11	14
80	·1736	1719	1702	1685	1668	1650	1633	1616	1599	1582	3	6	9	12	14
81	·1564	1547	1530	1513	1495	1478	1461	1444	1426	1409	3	6	9	12	14
82	·1392	1374	1357	1340	1323	1305	1288	1271	1253	1236	3	6	9	12	14
83	·1219	1201	1184	1167	1149	1132	1115	1097	1080	1063	3	6	9	12	14
84	·1045	1028	1011	0993	0976	0958	0941	0924	0906	0889	3	6	9	12	14
85	·0872	0854	0837	0819	0802	0785	0767	0750	0732	0715	3	6	9	12	15
86	·0698	0680	0663	0645	0628	0610	0593	0576	0558	0541	3	6	9	12	15
87	·0523	0506	0488	0471	0454	0436	0419	0401	0384	0366	3	6	9	12	15
88	·0349	0332	0314	0297	0279	0262	0244	0227	0209	0192	3	6	9	12	15
89	·0175	0157	0140	0122	0105	0087	0070	0052	0035	0017	3	6	9	12	15
90	·0000														

NATURAL TANGENTS

Degrees	0' 0°·0	6' 0°·1	12' 0°·2	18' 0°·3	24' 0°·4	30' 0°·5	36' 0°·6	42' 0°·7	48' 0°·8	54' 0°·9	Mean Differences 1	2	3	4	5
0	·0000	0017	0035	0052	0070	0087	0105	0122	0140	0157	3	6	9	12	15
1	·0175	0192	0209	0227	0244	0262	0279	0297	0314	0332	3	6	9	12	15
2	·0349	0367	0384	0402	0419	0437	0454	0472	0489	0507	3	6	9	12	15
3	·0524	0542	0559	0577	0594	0612	0629	0647	0664	0682	3	6	9	12	15
4	·0699	0717	0734	0752	0769	0787	0805	0822	0840	0857	3	6	9	12	15
5	·0875	0892	0910	0928	0945	0963	0981	0998	1016	1033	3	6	9	12	15
6	·1051	1069	1086	1104	1122	1139	1157	1175	1192	1210	3	6	9	12	15
7	·1228	1246	1263	1281	1299	1317	1334	1352	1370	1388	3	6	9	12	15
8	·1405	1423	1441	1459	1477	1495	1512	1530	1548	1566	3	6	9	12	15
9	·1584	1602	1620	1638	1655	1673	1691	1709	1727	1745	3	6	9	12	15
10	·1763	1781	1799	1817	1835	1853	1871	1890	1908	1926	3	6	9	12	15
11	·1944	1962	1980	1998	2016	2035	2053	2071	2089	2107	3	6	9	12	15
12	·2126	2144	2162	2180	2199	2217	2235	2254	2272	2290	3	6	9	12	15
13	·2309	2327	2345	2364	2382	2401	2419	2438	2456	2475	3	6	9	12	15
14	·2493	2512	2530	2549	2568	2586	2605	2623	2642	2661	3	6	9	12	16
15	·2679	2698	2717	2736	2754	2773	2792	2811	2830	2849	3	6	9	13	16
16	·2867	2886	2905	2924	2943	2962	2981	3000	3019	3038	3	6	9	13	16
17	·3057	3076	3096	3115	3134	3153	3172	3191	3211	3230	3	6	10	13	16
18	·3249	3269	3288	3307	3327	3346	3365	3385	3404	3424	3	6	10	13	16
19	·3443	3463	3482	3502	3522	3541	3561	3581	3600	3620	3	7	10	13	16
20	·3640	3659	3679	3699	3719	3739	3759	3779	3799	3819	3	7	10	13	17
21	·3839	3859	3879	3899	3919	3939	3959	3979	4000	4020	3	7	10	13	17
22	·4040	4061	4081	4101	4122	4142	4163	4183	4204	4224	3	7	10	14	17
23	·4245	4265	4286	4307	4327	4348	4369	4390	4411	4431	3	7	10	14	17
24	·4452	4473	4494	4515	4536	4557	4578	4599	4621	4642	4	7	11	14	18
25	·4663	4684	4706	4727	4748	4770	4791	4813	4834	4856	4	7	11	14	18
26	·4877	4899	4921	4942	4964	4986	5008	5029	5051	5073	4	7	11	15	18
27	·5095	5117	5139	5161	5184	5206	5228	5250	5272	5295	4	7	11	15	18
28	·5317	5340	5362	5384	5407	5430	5452	5475	5498	5520	4	8	11	15	19
29	·5543	5566	5589	5612	5635	5658	5681	5704	5727	5750	4	8	12	15	19
30	·5774	5797	5820	5844	5867	5890	5914	5938	5961	5985	4	8	12	16	20
31	·6009	6032	6056	6080	6104	6128	6152	6176	6200	6224	4	8	12	16	20
32	·6249	6273	6297	6322	6346	6371	6395	6420	6445	6469	4	8	12	16	20
33	·6494	6519	6544	6569	6594	6619	6644	6669	6694	6720	4	8	13	17	21
34	·6745	6771	6796	6822	6847	6873	6899	6924	6950	6976	4	9	13	17	21
35	·7002	7028	7054	7080	7107	7133	7159	7186	7212	7239	4	9	13	18	22
36	·7265	7292	7319	7346	7373	7400	7427	7454	7481	7508	5	9	14	18	23
37	·7536	7563	7590	7618	7646	7673	7701	7729	7757	7785	5	9	14	18	23
38	·7813	7841	7869	7898	7926	7954	7983	8012	8040	8069	5	9	14	19	24
39	·8098	8127	8156	8185	8214	8243	8273	8302	8332	8361	5	10	15	20	24
40	·8391	8421	8451	8481	8511	8541	8571	8601	8632	8662	5	10	15	20	25
41	·8693	8724	8754	8785	8816	8847	8878	8910	8941	8972	5	10	16	21	26
42	·9004	9036	9067	9099	9131	9163	9195	9228	9260	9293	5	11	16	21	27
43	·9325	9358	9391	9424	9457	9490	9523	9556	9590	9623	6	11	17	22	28
44	·9657	9691	9725	9759	9793	9827	9861	9896	9930	9965	6	11	17	23	29

NATURAL TANGENTS

Degrees	0' 0°·0	6' 0°·1	12' 0°·2	18' 0°·3	24' 0°·4	30' 0°·5	36' 0°·6	42' 0°·7	48' 0°·8	54' 0°·9	Mean Differences 1	2	3	4	5
45	1·0000	0035	0070	0105	0141	0176	0212	0247	0283	0319	6	12	18	24	30
46	1·0355	0392	0428	0464	0501	0538	0575	0612	0649	0686	6	12	18	25	31
47	1·0724	0761	0799	0837	0875	0913	0951	0990	1028	1067	6	13	19	25	32
48	1·1106	1145	1184	1224	1263	1303	1343	1383	1423	1463	7	13	20	27	33
49	1·1504	1544	1585	1626	1667	1708	1750	1792	1833	1875	7	14	21	28	34
50	1·1918	1960	2002	2045	2088	2131	2174	2218	2261	2305	7	14	22	29	36
51	1·2349	2393	2437	2482	2527	2572	2617	2662	2708	2753	8	15	23	30	38
52	1·2799	2846	2892	2938	2985	3032	3079	3127	3175	3222	8	16	24	31	39
53	1·3270	3319	3367	3416	3465	3514	3564	3613	3663	3713	8	16	25	33	41
54	1·3764	3814	3865	3916	3968	4019	4071	4124	4176	4229	9	17	26	34	43
55	1·4281	4335	4388	4442	4496	4550	4605	4659	4715	4770	9	18	27	36	45
56	1·4826	4882	4938	4994	5051	5108	5166	5224	5282	5340	10	19	29	38	48
57	1·5399	5458	5517	5577	5637	5697	5757	5818	5880	5941	10	20	30	40	50
58	1·6003	6066	6128	6191	6255	6319	6383	6447	6512	6577	11	21	32	43	53
59	1·6643	6709	6775	6842	6909	6977	7045	7113	7182	7251	11	23	34	45	56
60	1·7321	7391	7461	7532	7603	7675	7747	7820	7893	7966	12	24	36	48	60
61	1·8040	8115	8190	8265	8341	8418	8495	8572	8650	8728	13	26	38	51	64
62	1·8807	8887	8967	9047	9128	9210	9292	9375	9458	9542	14	27	41	55	68
63	1·9626	9711	9797	9883	9970	2·0057	2·0145	2·0233	2·0323	2·0413	15	29	44	58	73
64	2·0503	0594	0686	0778	0872	0965	1060	1155	1251	1348	16	31	47	63	78
65	2·1445	1543	1642	1742	1842	1943	2045	2148	2251	2355	17	34	51	68	85
66	2·2460	2566	2673	2781	2889	2998	3109	3220	3332	3445	18	37	55	73	92
67	2·3559	3673	3789	3906	4023	4142	4262	4383	4504	4627	20	40	60	79	99
68	2·4751	4876	5002	5129	5257	5386	5517	5649	5782	5916	22	43	65	87	108
69	2·6051	6187	6325	6464	6605	6746	6889	7034	7179	7326	24	47	71	95	119
70	2·7475	7625	7776	7929	8083	8239	8397	8556	8716	8878	26	52	78	104	131
71	2·9042	9208	9375	9544	9714	9887	3·0061	3·0237	3·0415	3·0595	29	58	87	116	145
72	3·0777	0961	1146	1334	1524	1716	1910	2106	2305	2506	32	64	96	129	161
73	3·2709	2914	3122	3332	3544	3759	3977	4197	4420	4646	36	72	108	144	180
74	3·4874	5105	5339	5576	5816	6059	6305	6554	6806	7062	41	81	122	163	204
75	3·7321	7583	7848	8118	8391	8667	8947	9232	9520	9812	46	93	139	186	232
76	4·0108	0408	0713	1022	1335	1653	1976	2303	2635	2972	53	107	160	213	267
77	4·3315	3662	4015	4374	4737	5107	5483	5864	6252	6646					
78	4·7046	7453	7867	8288	8716	9152	9594	5·0045	5·0504	5·0970	Mean differences cease				
79	5·1446	1929	2422	2924	3435	3955	4486	5026	5578	6140	to be sufficiently accurate.				
80	5·6713	7297	7894	8502	9124	9758	6·0405	6·1066	6·1742	6·2432					
81	6·3138	3859	4596	5350	6122	6912	7720	8548	9395	7·0264					
82	7·1154	2066	3002	3962	4947	5958	6996	8062	9158	8·0285					
83	8·1443	2636	3863	5126	6427	7769	9152	9·0579	9·2052	9·3572					
84	9·5144	9·677	9·845	10·02	10·20	10·39	10·58	10·78	10·99	11·20					
85	11·43	11·66	11·91	12·16	12·43	12·71	13·00	13·30	13·62	13·95					
86	14·30	14·67	15·06	15·46	15·89	16·35	16·83	17·34	17·89	18·46					
87	19·08	19·74	20·45	21·20	22·02	22·90	23·86	24·90	26·03	27·27					
88	28·64	30·14	31·82	33·69	35·80	38·19	40·92	44·07	47·74	52·08					
89	57·29	63·66	71·62	81·85	95·49	114·6	143·2	191·0	286·5	573·0					
90	∞														

EXPONENTIAL FUNCTIONS

x	e^x	e^{-x}	x	e^x	e^{-x}
·02	1·0202	·9802	1·0	2·7183	·3679
·04	1·0408	·9608	1·1	3·0042	·3329
·06	1·0618	·9418	1·2	3·3201	·3012
·08	1·0833	·9231	1·3	3·6693	·2725
·10	1·1052	·9048	1·4	4·0552	·2466
·11	1·1163	·8958	1·5	4·4817	·2231
·12	1·1275	·8869	1·6	4·9530	·2019
·13	1·1388	·8781	1·7	5·4739	·1827
·14	1·1503	·8694	1·8	6·0497	·1653
·15	1·1618	·8607	1·9	6·6859	·1496
·16	1·1735	·8521	2·0	7·3891	·1353
·17	1·1853	·8437	2·1	8·1662	·1225
·18	1·1972	·8353	2·2	9·0250	·1108
·19	1·2092	·8270	2·3	9·9742	·1003
·20	1·2214	·8187	2·4	11·023	·0907
·21	1·2337	·8106	2·5	12·182	·0821
·22	1·2461	·8025	2·6	13·464	·0743
·23	1·2586	·7945	2·7	14·880	·0672
·24	1·2712	·7866	2·8	16·445	·0608
·25	1·2840	·7788	2·9	18·174	·0550
·26	1·2969	·7711	3·0	20·085	·0498
·27	1·3100	·7634	3·1	22·198	·0450
·28	1·3231	·7558	3·2	24·532	·0408
·29	1·3364	·7483	3·3	27·113	·0369
·30	1·3499	·7408	3·4	29·964	·0334
·31	1·3634	·7335	3·5	33·115	·0302
·32	1·3771	·7261	3·6	36·598	·0273
·33	1·3910	·7189	3·7	40·447	·0247
·34	1·4050	·7118	3·8	44·701	·0224
·35	1·4191	·7047	3·9	49·402	·0202
·36	1·4333	·6977	4·0	54·598	·0183
·37	1·4477	·6907	4·1	60·340	·0166
·38	1·4623	·6839	4·2	66·686	·0150
·39	1·4770	·6771	4·3	73·700	·0136
·40	1·4918	·6703	4·4	81·451	·0123
·41	1·5068	·6636	4·5	90·017	·0111
·42	1·5220	·6570	4·6	99·484	·0100
·43	1·5373	·6505	4·7	109·95	·00910
·44	1·5527	·6440	4·8	121·51	·00823
·45	1·5683	·6376	4·9	134·29	·00745
·46	1·5841	·6313	5·0	148·41	·00674
·47	1·6000	·6250	5·1	164·02	·00610
·48	1·6161	·6188	5·2	181·27	·00552
·49	1·6323	·6126	5·3	200·34	·00499
·50	1·6487	·6065	5·4	221·41	·00452
.60	1·8221	·5488	5·5	244·69	·00409
·70	2·0138	·4966	5·6	270·43	·00370
·80	2·2255	·4493	5·7	298·87	·00335
·90	2·4596	·4066	5·8	330·30	·00303
			5·9	365·04	·00274
			6·0	403·43	·00248

SUBJECT INDEX